Fairchild Bros. New York

Fairchild's Hand-Book of the digestive Ferments

As Remedies, per se, as surgical Solvents and in the Peptonisation of Milk

and other Foods for the Sick

Fairchild Bros. New York

Fairchild's Hand-Book of the digestive Ferments
As Remedies, per se, as surgical Solvents and in the Peptonisation of Milk and other Foods for the Sick

ISBN/EAN: 9783337184049

Printed in Europe, USA, Canada, Australia, Japan

Cover: Foto ©berggeist007 / pixelio.de

More available books at **www.hansebooks.com**

Fairchild's Hand-Book

OF THE

Digestive Ferments

AS REMEDIES, *PER SE*

AS SURGICAL SOLVENTS

AND IN THE

PEPTONISATION OF MILK AND OTHER
FOODS FOR THE SICK

AND FOR

THE MODIFICATION OF COW'S MILK TO THE
STANDARD OF HUMAN MILK

BY THE FAIRCHILD PROCESS

FAIRCHILD BROS. & FOSTER

NEW YORK

82 & 84 FULTON STREET

LONDON

SNOW HILL BUILDINGS

WESTERN DEPOT

110 RANDOLPH STREET, CHICAGO

FAIRCHILD'S PREPARATIONS.

Pepsin in Scales,
Pepsin in Powder,
Essence of Pepsine,
Saccharated Pepsin,
Glycerinum Pepticum,
Extractum Pancreatis,
Diastasic Essence of Pancreas,
Peptonising Tubes,
Peptogenic Milk Powder,
Panopepton,
Pancreatic Tablets,
Compound Pancreatic Tablets,
Pepsin and Extract Pancreatis Tablets,
Pepsin and Bismuth Tablets,
Pepsin, Bismuth and Pancreatic Tablets,
Pepsin, Bismuth and Nux Vom. Tablets,
Pepsin and Diastase Tablets,
Peptonate of Iron Tablets,
Compound Ox Gall Tablets,
Ferroglobin Tablets,
Trypsin.

HE Various Uses of the Fairchild Digestive Ferments described in the pages of this pamphlet, afford striking evidence of the great progress made in their development and practical application, in the past decade. The manufacture of the Pepsin in Scales and the Extractum Pancreatis was undertaken with an enthusiasm based upon our conviction of the valuable properties and possibilities of the gastric and pancreatic ferments as therapeutic and as peptonising agents, and of the inferior and often useless character of the pancreatines and pepsins of the market.

Such was the demand made upon our time and upon our resources, that we soon found it expedient to relinquish the general drug business in order to devote ourselves exclusively to the digestive ferments. In this work we have sought to develop and perfect the means and the methods of their employment in every useful direction ; and we have been happy to see them assume so important a place in practical medicine ; a place won on the basis of properties actually demonstrated and applied.

The modern application of the digestive ferments by the Fairchild process to the predigestion of food for the sick may be considered as direct, natural and scientific a development of the resources lying at our hands, as the art of cookery itself, by which we seek to perfect the adaptation of food stuffs to the needs of mankind.

In no direction has a digestive ferment been of more positive benefit and radical improvement than in the preparation of infant food. The use of a digestive ferment as the essential factor in a method for the preparation of an artificial human milk, was first suggested and brought to a practical form by us in the Peptogenic Milk Powder.

So strong is our conviction of the soundness of the principles upon which this method is based, of the accuracy with which it is brought into practice, so conclusive the evidence of its beneficent results in actual use, that we feel constrained to urge its claims to consideration as a means of obtaining a complete and exclusive substitute for breast milk.

As surgical solvents the digestive ferments justify serious attention, for they are potent, are painless in their work and invade and liquefy dead tissue straight down to living ones, where their action ends abruptly ; they impart moreover a distinct stimulus to the healing process.

The Fairchild preparations are each and all the result of persistent, careful, special work, and we believe them to be not only the original, but the best, for all the purposes designed. They have long been conceded to be the standard.

We desire to take this occasion to again express our appreciation of the confidence and recognition so generously accorded to our efforts by the medical profession. It is no less a source of pleasure to us that so many of the best pharmacists find our products worthy of their careful and discriminating preference.

FAIRCHILD BROS. & FOSTER.

Revised Edition, April, 1893.

CONTENTS

DIGESTIVE FERMENTS—GENERAL CHARACTERISTICS.
Action of ; Limit of energy ; Changes of alimentary substances by digestion ; In a dry form, should not be hygroscopic ; Dry ferments compatible with substances which would injure them in solution ; Dry compounds as subject to assay as the separate ferments from which prepared ; Relation to temperature ; In solution ; Incompatibility of solutions of mixed ferments..........pp. 10-16

ALCOHOL and the digestive ferments ; Action of strong and diluted ; Value as a preservative...pp. 16-18

INHIBITANTS. The influence of drugs and dietary substances upon the process of artificial digestion ; The relation of test tube experiments to the conditions of body digestion ; necessity of distinguishing between substances which retard the process of artificial digestion and substances which destroy the ferment..pp. 19-22

INCOMPATIBLES. Substances and conditions which destroy the digestive ferments ; Pepsin and bismuth in solution..............pp. 23-25

ANTISEPTICS ... .pp. 25-26

"JUMBLES." Character of the commercial digestive compounds, their good effects due to pepsin and acid.....................p. 26

VEGETABLE FERMENT. Its inferiority in comparison with pepsin in acid media and pancreatic extract in alkaline or neutral...pp. 27-28

GASTRIC FERMENTS. Pepsin, its action upon albumen ; Peptic peptone ; Pepsin inert in alkaline solution ; Pepsin and soda ; Practical uses of pepsin............................... ......pp. 28-30

MILK CURDLING FERMENT of the Gastric Juice. Its action upon caseine ; Significance in the digestive process..........pp. 30-32

PANCREATIC FERMENTS. Trypsin, Action of ; Tryptic peptones ; Pancreatic diastase ; Identity of diastase from all sources in properties and action ; Uses of Emulsive Ferment ; Curdling ferment
pp. 32-38

DOSAGE of digestive ferments... ....pp. 35-36

7

FAIRCHILD PREPARATIONS of the Digestive Ferments ; Uniformity and reliability ; Their repute based upon actual demonstrated properties ; Commercial imitations and substitutes therefor.....pp. 38-42

PEPSIN IN SCALES AND POWDER. The most active and desirable pepsin for administering in a dry form ; Pepsin in scales first introduced by Fairchild ; Absolute permanency............pp. 42-43

PEPSIN SACCHARATED. Feeble power of ; Only officinal dry pepsin preparation..p. 43

GLYCERINUM PEPTICUM. A pure glycerin extract from the gastric membrane, free from alcohol, antiseptics, sugar or flavoring ; The best soluble form of pepsin........................pp. 43-44

ESSENCE OF PEPSINE, Fairchild's. Obtained by direct maceration from the fresh calf rennet ; Special value as a remedy in disorders of infancy, and dyspepsia of adults ; As a means of administering drugs which disturb the digestive functions and impair the appetite ; As a practical rennet agent..... ....pp. 45-47

MEDICATED JUNKET. Suggested by Dr. Delavan ; Milk-curd made with Fairchild's Essence containing Potassium Iodide in solution, the curd holding iodide suspended in a very agreeable form..pp. 47-48

PEPSIN TESTING..pp. 48-52

PANCREATIC PREPARATIONS. Extractum Pancreatis containing all the ferments of the pancreas in an active and available form ; As a remedy *per se ;* As a diastasic, proteolytic and emulsifying agent ; Its special value in intestinal indigestion..............pp. 52-57

TRYPSIN, Fairchild's. As a solvent of diphtheritic membrane....p. 57

DIASTASIC ESSENCE OF PANCREAS, especially for the digestion of farinaceous foods..p. 58

PEPTONISING TUBES, for the preparation of Peptonised Milk, etc..p. 59

DIRECTION SLIPS. For prescribing Peptonised Milk, etc........p. 59

PEPTOGENIC MILK POWDER........ ....p. 60

TESTS for pancreatic preparations......................pp. 60-62

DIGESTIVE TABLETS, Fairchild's. Pepsin Tablets ; Pepsin and Bismuth Tablets ; Pepsin, Bismuth and Pancreatic Tablets ; Pepsin and Pancreatine Tablets ; Pepsin and Diastase Tablets ; Pepsin, Bismuth and Nux Vomica Tablets ; Compound Ox Gall Tablets ; Pancreatic Tablets ; Compound Pancreatic Tablets ; Peptonate of Iron Tablets ; Ferroglobin Tablets ....pp. 62-67

PEPTONISING PROCESS. Simplicity, economy and practicability of ; Use of soda in ; Reasons for diluting milk in..........pp. 67-73

USES OF PEPTONISED FOODS. In Typhoid Fever; pneumonia; gastric ulcer; acute dysentery; diabetes; tuberculosis; chronic diarrhœa; gastric catarrh; value as exclusive diet even in active life.pp. 73-76

PEPTONISED MILK. Ideal food for the sick................pp. 77-78

NUTRITIVE ENEMATA. Milk, beef, etc......... ....pp. 78-79

PANOPEPTON—Bread and Beef Peptone; a properly digested, complete nutrient................................ ....pp. 79 82

SURGICAL USE of the Digestive Ferments. As solvents for false fibrinous membrane, coagula, muco-pus, necrotic and carious bone; applied in aural cavity, urethra, bladder, etc............pp. 83-90

PEPTOGENIC MILK POWDER. For the preparation of humanised milk, Identical with normal human milk in physical, chemical and physiological properties; Rationale of the process; Agency of the digestive ferment as an innocent, practical, and only known means of converting caseine into the soluble form characteristic of the albuminoids of human milk.......... ....pp. 91-94

INFANT FOODS. Only practical point of inquiry; How do they compare with breast milk when prepared for the nursing bottle; Fresh cows' milk only practical basis for making an infant food; Milk Foods, etc.; Impossibility of drying pure milk....... ....pp. 94-95

COWS' MILK. Proven inherently indigestible for an infant's stomach; Common methods of preparing it for infants; Liebig's Food; Farinaceous Food............................. ....pp. 95-97

COMPARATIVE COMPOSITION OF COWS' AND HUMAN MILK. Difference in their physical characters, digestibility behavior with gastric juice directly due to difference in their albuminoids; significant difference also in proportion and quantity of nutritive materials pp. 97-98

USE OF PEPTOGENIC POWDER. Includes the preparation of an exact quantitative imitation of human milk, exact qualitative change of albuminoids and subsequent destruction of the ferment...pp. 98-99

DIRECTIONS FOR "HUMANISED MILK"; in health and in feeble digestion..pp. 99-100

COMPOSITION OF "HUMANISED MILK." Remarkably like average breast milk in chemical constitution, reaction, density, color, taste and in behavior under all conditions....................p. 100

DIGESTIBILITY OF "HUMANISED MILK"; Not unnaturally easy of digestion; as digestible as mothers' milk; adapted for feeble digestion by increasing the pre-digestion of the caseine. ...pp. 101-102

CHOLERA INFANTUM. Whey as a temporary food in......pp 102 104

How Long Should Infant be fed upon humanised milk ; Humanised milk the only food suitable during entire nursing period.p. 104

How to Wean the bottle-fed baby....................pp. 104-105

As a Partial Substitute for Breast Milk ; Humanised milk most successful *partial* food, because so like breast milk.........p. 105

No Special Effect upon the Bowels from " humanised milk "
pp. 105-107

Changing the Food ; Evil of going from one food to another without definite knowledge or basis of selection ....pp. 107-109

Rich Milk from one cow ; Cream : Temperature of the water bath ; Milk tastes bitter ; Milk curdled when boiled........pp. 109-112

Condensed Milk ....pp. 112-113

Sterilised Milk.   Character of ; Effect of sterilising process ; less digestible and far less nutritious than fresh milk.......pp. 113-115

Comparative Analyses average of 80 samples of woman's milk and of "humanised milk," by Dr. Albert R. Leeds ....p. 116

Fairchild's Practical Recipes ; For peptonising food for the sick ; Nutritive value of milk compared with beef tea, extracts of beef, etc ....pp. 117-118

Peptonised Milk. Warm process ; Cold process ; Hot, as a beverage ; Effervescent ; Special for jellies, punches, etc.; Punch ; Lemonade ; Peptonised milk gruel; Peptonised porridge ; Beef ; Oysters ; Junket and whey with Fairchild's Essence of Pepsine ; Partial digestion of farinaceous foods at the table.....pp. 119-125

List of Fairchild's Preparations ....p. 126

DIGESTIVE FERMENTS.

In the digestive ferments, we have to deal with an entirely distinct class of agents, bearing little or no analogy to drugs and chemicals. They are not known to exert any action in the body other than that concerned in the conversion of alimentary substances into soluble and absorbable forms. By this action alone we know them and can determine their presence.

No digestive ferment has been absolutely isolated, consequently the chemical constitution of these principles is yet a matter of conjecture. We do not know how they perform their marvellous work, nor the exact chemical formula of the various derivatives of digestion. These limitations to our knowledge of the digestive ferments do not impose any limitations upon our practical use of them. For we are able to extract them from the digestive juices or secreting glands and to preserve them indefinitely as reliable agents of the materia medica. We know well the conditions under which they act, what is unfavorable to their action, what is directly destructive to them.

We can as unerringly detect the presence of pepsin or diastase as that of morphia or strychnia. We can readily ascertain the digestive power or value of any given product. The physical changes of alimentary bodies under artificial digestion are so characteristic, so apparent to sight and taste that they afford convenient and familiar evidence by which the peptonising process may be as readily regulated as that of cooking.

It is true, the digestive ferments are profoundly sensitive to influences which have little or no effect upon the medicinal properties of drugs and chemicals. Furthermore, there have been many fallacious statements and theories concerning the digestive ferments, and the immense array of experiments, with the theoretical discussions thereon, have also tended to give an undue impression of peculiar difficulties attending their practical use. The truth is that the digestive ferments may be prescribed with the same certainty, with as definite and well grounded anticipation, with as little difficulty as regards incompatibility, as in the use of drugs and chemicals.

The digestive ferments find their entire use in three distinct directions. First, as remedies *per se*, as aids to the digestive process within the body. Here the main concern of the physician is to avoid prescribing the digestive ferments with substances which injure them ; there are but few of these at all likely to be prescribed with the digestive ferments and they are in the discussion of "Incompatibles" conveniently summarised for reference. In the therapeutic use of the digestive ferments the influence of drugs, etc., on the process of digestion is apt also to be considered. This question is, however, without practical bearing here. For the action of the so-called "inhibitants" is only ascertained in the test tube where the action of each ferment is clogged even by the products of digestion and retarded by substances which in the stomach would have no influence whatever. Under the subject of "inhibitants," the relation of these experiments to the conditions of body digestion and the deductions to be drawn therefrom are fully discussed.

In the artificial digestion of foods, the relation of the digestive ferments to temperature is such that simply by its regulation, we may obtain their energetic action, may hold the ferments in a latent form, or instantly and permanently check action at any given stage, as described in the peptonising process.

For the solution of morbid tissue, we have but to employ the special ferment indicated, in its proper vehicle, and remove by irrigation both the ferment and the dissolved matter. The certainty with which the Fairchild preparations of the digestive ferments act, either upon alimentary bodies, or morbid tissues, affords sufficient proof that the digestive ferments are not necessarily variable or unreliable agents.

Our work with the digestive ferments has been of that practical character involved in the production of these organic principles in the most active and best form ; and in the invention and development of preparations and processes for their application. In the following pages, we have sought to present the salient facts concerning the digestive ferments in the whole range of their relations to the conditions and agents with which they are practically brought into contact, and to describe the proper methods for their employment as therapeutic and peptonising agents.

DIGESTIVE FERMENTS.

GENERAL CHARACTERISTICS.

The digestive ferments belong to the class of soluble unorganised ferments, possessing no power of self-nutrition or self-multiplication. They differ entirely in their mode of action from living ferments, such as yeast or bacteria. Their action is further unaccompanied by the phenomena ordinarily associated with fermentation. They may be described as agents capable of setting up between substances, under conditions of moderate temperature, a chemical action of which these substances are incapable without the intervention of the ferment. The digestive ferments probably belong to the proteid class, or are closely related thereto. At present in the most active form in which they are practically obtained, they are found to correspond in behavior and constitution to proteid bodies. They are all soluble in water, and by simple infusion of the fresh gland or the secreting membrane, we may obtain active solutions which exhibit in the proper media all the behavior of the natural juices. They resemble in a degree all ferments in their energy, in the minute proportion required in the conversion of alimentary substances.

Stress has been laid upon the fact that a ferment after having performed a certain amount of digestion may be recovered and made to repeat its work, and it has been by some writers assumed that the power of a ferment is limitless. Even if this theory were true, it has no bearing upon ordinary operations with a digestive ferment upon its correlated substance, to determine its extent of energy under definite conditions or to utilise its force. Nor is it of any significance as to the role of the ferment in the normal process of digestion. It is of no bearing whatever in the use of the ferment for any practical purpose, for artificial solvents of alimentary substances or of morbid tissues, etc., or as aids to digestion.

It is, however, a fact that the digestive ferment has a definite, ascertainable limit of energy ; its power is used up just in proportion to the work done.

When under certain conditions favorable to digestion, with arbitrary proportions of alimentary substance and media a point is found at which a given amount of ferment leaves a large excess of substance unaffected, it is because the ferment has lost all its power. Thus in a series of experiments with increasing ratio of substance to ferment, we ascertain the relative as well as actual power of pepsin or diastase for instance.

The characteristic action of the digestive ferment is

the conversion of alimentary substances into the peculiar soluble form essential to their absorption. But the action of the ferments is not restricted to alimentary bodies ; the proteolytic ferments, both of the stomach and the pancreas gland, are capable of digesting albuminous or fibrinous substances, such as false membranes, coagula, etc.

It is under physiological conditions that the ferments produce changes which can otherwise only be approximated under high temperature and chemical reagents ; as, for instance, in the making of peptones or glucose by prolonged boiling with acid. The change which all alimentary bodies undergo during digestion profoundly affects their physical properties, and consequently their susceptibility to osmosis, whilst their chemical composition is but slightly altered. In order to distinguish the digestive ferments from living or organised ferments, Kuhne proposed to call them enzymes, and further it has been proposed to distinguish their action as enzymatic, in contrast to true fermentation. But these terms have but little practical recognition, and the distinctions between digestive ferments and living, yeast, or germ ferments, are now so well understood that the use of the term "digestive ferments" really leads to little confusion.

In a dry form the digestive ferments permanently retain their properties. For inasmuch as water is essential to the action of the digestive ferment so the presence of water is essential to its reaction with any other substance. Moisture and heat are favorable to their decomposition. An essential quality of dry products of the digestive ferments is, that they shall not be prone to absorb moisture—shall not be hygroscopic. They should never be prescribed in combination with deliquescent salts, peptone, etc. A digestive ferment may properly be combined in a dry form with substances with which it should not be brought into contact in solution. Thus dry pepsin will not be injured by contact with soda bicarbonate.

Obviously in a dry form, the digestive ferments are without action towards each other. Therefore, a mixture of these ferments should retain, and under proper conditions, must exhibit the behavior characteristic of each one of the ferments contained. For with several ferments placed in a digesting mass, those under conditions unfavorable to action have no possible interference with the action of the particular ferment for which the conditions are appropriate. A digestive compound is therefore, in every particular, as subject to assay as the separate ferments from which it is prepared. For instance, if a powder contains "pepsin, pancreatine and diastase," it should in acidulated water, give all the results of the contained quantity of pepsin, in an alkaline medium it should digest fibrin or milk, and in a neutral or alkaline solution liquefy gelatinous starch. Whatever theory or opinion may be held concerning the propriety of such combinations, it must certainly be obvious that their value as *digestive* agents must as much depend upon the possession of the digestive properties of the various ferments, as the value of a preparation of pepsin or of pancreatic extract is measured by the degree in which it exhibits peptic or pancreatic activity.

The digestive ferments are inert but not injured, at a low temperature. They bear prolonged exposure to the freezing-point without becoming impaired. In solution, at the ordinary temperature of a room, 70° F., they act slowly, favorably at the temperature of the body, and increasingly up to about 130° F., when as the temperature rises they sharply diminish in activity until at about 160° F., they are quite destroyed.

Pepsin is active only with acid ; pancreatic ferments in neutral, alkaline and feebly acid solutions.

It is impossible to prepare a menstruum suitable for the solution and preservation of mixed ferments of the pancreas and the stomach. If we mix active solutions of the stomach and of the pancreas and test the mixture after it has been set aside at the ordinary temperature of the room for a few days, it will be found that the mixture no longer represents all the digestive ferments as contained in the original solutions. If the reaction of the solution of the mixed ferments has been neutral or alkaline, the pepsin will have been destroyed ; if acid, the pancreatic ferments will have lost their properties. It may be said, therefore, without qualification, that the whole class of fluid mixtures of gastric and pancreatic ferments are un-scientific, and invariably devoid of most of the ferments they purport to contain.

ALCOHOL AND THE DIGESTIVE FERMENTS.

The digestive ferments vary so little in their behavior with alcohol, that they may all be said to bear a common relation to it. They are insoluble in alcohol, soluble in diluted alcohol and precipitated from solution by alcohol in excess.

To effectually employ alcohol as a precipitant of the ferments, they must be held in a concentrated solution, to which the stronger alcohol must be added in such a volume as to give the largest practicable percentage of absolute alcohol. The ferments so recovered, may again be re-dissolved in water.

That alcohol does not destroy the ferments, may be seen in the method commonly employed by physiological chemists, in extracting the ferments in a pure solution convenient for experimental purposes. The mucous membrane or gland is first exposed to alcohol which washes, hardens and dehydrates it ; then to a solvent (glycerin preferably) which takes up the ferment largely free from inert extractives, coloring matter, etc.

But notwithstandidg the fact that alcohol is thus recommended in leading works on physiology, we have been unable to convince ourselves that strong alcohol does not exert a direct injurious action on the ferments. The alcohol separated ferment does not exhibit the activity which it should theoretically possess, calculated upon the degree of isolation and the known assayed ferment power of the original infusion of the gland. But the degree of activity suitable for the physiological chemist, who simply requires solutions of the ferments capable of exhibiting the characteristic reaction, is no doubt far inferior to the standard attained by the manufacturing chemist in applying ferments to practical purposes.

From a pharmaceutical standpoint alcohol bears in some respects the same relation to the digestive ferments as it does to many drugs. A watery infusion from the stomach, like all other infusions of organic substances, will be soon decomposed and the ferments therein will lose all activity unless there is some preservative added. A hydro-alcoholic menstruum serves as useful a purpose in extracting the digestive ferments as it does in extracting the active principle of drugs. The first desideratum of fluid preparations is that they should present effective doses in a moderate volume. With a menstruum containing say 15 to 20 per cent. of pure spirit, all the ferments may be extracted and preserved in an effective form for medicinal purposes or for use in the artificial digestion of food. It is not by any means a sufficient cause for the rejection of this class of preparations, merely because they contain alcohol up to say 20 per cent. of volume. In the percentage sufficient as a preservative, alcohol does not necessarily injure the ferments or render them inert. As present in this proportion it becomes an insignificant factor in so far as it affects the value of a digestive fluid, owing to the dilution it will receive in practical uses.

We must require of such a preparation as the essential ground of its employment, that it shall exhibit actual digestive power, the characteristic action of the ferment which it purports to represent, when submitted to the identical conditions used in assaying the dry ferments themselves.

The solutions or liquid extracts from the pancreas are objectionable and inferior to the dry Extractum Pancreatis, not because of the 20 p. c. of alcohol, but because of the tendency of these solutions to precipitate, to undergo deterioration owing to the large amount of organic matter they contain. The diastasic power is especially variable and weak, and tends to constantly diminish. These solutions further impart their peculiar repulsive taste to foods, milk, gruel, etc., and consequently they have found little usage, and now are entirely superseded by the Extractum Pancreatis.

The question of the influence which alcohol exerts upon the artificial process of digestion, its bearing upon the proper use of alcohol in fluid preparations of the digestive ferments, will be discussed when we come to consider the subject of inhibitants or substances which retard artificial digestion in the flask or test tube.

That many of the class of preparations, such as wines, elixirs, etc., are inefficient, is not due to the presence of alcohol, but for the reason that they have not been properly prepared, or have been made from commercial products originally deficient in digestive properties. For this class of preparations of the digestive ferments will be found to vary very much, as do galenical preparations generally, according to the skill and technical knowledge exercised in their manufacture.

INHIBITANTS.

No question concerning the digestive ferments has been given more attention than the influence of medicinal and dietary substances upon the process of artificial digestion. It has been the subject of many experiments and raised many speculations. We have had elaborate tests, giving the exact observed degree of retardation exhibited by a great variety of drugs and chemicals, some of which would scarcely by any chance ever be mixed with a digestive ferment in practice ; also of the effect of alcohol, wine, spirits, beer, tea, cocoa, coffee, whey, sugar, common salt, etc. That various observers reach conflicting results and conclusions is due to the fact that no two employ digestive fluids of the same strength or follow precisely the same method in detail. Whilst these experiments are very interesting and attractive, the real point of inquiry must be to determine their practical bearing in medicine and pharmacy. There is a very necessary distinction to be drawn between the action of substances upon the ferments direct and upon the digestive process. It is of the greatest importance to the physician and the pharmacist to know the agents which destroy the ferment when brought into contact with it. But as to the practical significance of this whole class of experiments showing the retarding effect of substances upon the process of digestion, we must consider what relation or resemblance exists between the conditions in the test tube and in the living body. We must ask why and how these agents retard digestion and if they are likely to produce similar results when they are taken into the body. The common method of experiment is to take, say with pepsin for illustration, a definite amount of the ferment, albumen, acid and water up to an arbitrary volume, the proportions adjusted to produce a known amount of digestion in a definite time at blood

heat. This constitutes the control test. Into this mixture, in a series of tubes, are added the agents to be
tested and the effect upon digestion noted. These conditions in the test tube imitate those of the digestive
tract in temperature and in media ; they differ therefrom
in material points. In the test tubes, the very accuracy
of the proportions of the mixture, whilst essential to correct observation in experiments, in reality involves a great
fallacy. Water is essential to all physiological action ;
water is the only fluid in which and by which a digestive
ferment can act upon an alimentary substance. In the
stomach and intestinal canal there is not an arbitrary fixed
volume of liquid, which may be to a definite and known
degree altered by the addition of any substance. In the
normal digestive apparatus, the ferment may be said to
act in a current of water ; there is a constant secretion
of digestive juices during the entire period of action.
There is meanwhile a marked fluctuation in the reaction
and the composition of the digesting mass, owing to the
very complex nature of the substances of food and the
more or less definite chemical changes and combinations
formed therefrom. The products of digestion, the saline
constituents of food, are continuously absorbed in the
digestive tracts, leaving the digestive juice unhampered
in its work. In the test tube in the "control," the
first essential is a fixed volume of water. Now if in
another tube, we add a substance which reduces the proportion of water to any marked extent, we shall find, as
may only be anticipated, that we get less digestion. A
tube containing 80 per cent. of water and 20 per cent. of
alcohol or glycerin or sugar or peptone, will give less
result, not because these substances injure the ferment,
but because they cannot replace water, because of the
lessened value of the media for digestive action. Pure
glycerin exerts no injurious action upon a ferment, but

the ferment cannot transform albumen into peptone in glycerin. Again, we see in artificial digestive operations that when the fluid has become saturated with the products of digestion, the ferment can act no further. Not because peptone injures pepsin or maltose injures diastase, but because the water can take up no more and has no further power as a media for the ferment. Pure alcohol in excess is a precipitant of pepsin, of albumen and of peptone. About 10 per cent. absolute alcohol distinctly retards digestion in a test tube, but not because it is in this percentage injurious to the ferment. On the contrary, as already shown in the proportion of 15 to 20 per cent. it affords a most valued preservative of infusion of the ferments. Other substances retard digestion simply because they reduce or change the reaction of the media, as shown for example in peptic digestion by the fact that if the saturating powder of an added substance is compensated for by the addition of free acid to the percentage of the control, little or no retardation is found. The retarding influence of certain substances is modified by the strength of the digestive fluid—for instance, by the proportion of the pepsin to the albumen. If we take the utmost limit of albumen which a grain of pepsin can digest in several hours, say 2,000 grains, we shall find the digesting mass much more sensitive to salt, for instance, than one containing a grain of the same pepsin to 200 grains albumen ; the percentage of salt being the same in each case. This would seem to show that a powerful digestion is not affected like a feeble digestion—the retardation is relative not absolute. In the behavior of common salt in artificial peptic digestion, we have an illustration of the inadequacy of tests of "inhibitants" as guides to therapeutic uses of the digestive ferments or as explaining or approximating to the digestion in the body. Salt strongly retards the action of pepsin upon albumen in the test tube. It does not injure the ferment. On the contrary, it is a well known precipitant and preservative of pepsin. Salt inhibits digestion in a percentage which does not throw out the pepsin, nor affect the solvent action of water upon peptone, nor alter the reaction of the digesting mass. In view of these facts and of the universal use of salt as a condiment and antiseptic, we are at a loss to explain its retarding effects in artificial digestion and cannot believe it to exert any similar effect in the stomach.

The stomach, moreover, is endowed with the power of maintaining the physiological conditions essential to digestion. The ingestion of an alkali may neutralise morbific acids and provoke the secretion of the acids of digestion. Acids form various combinations with proteids and bases of food substances. Substances may, like alcohol, retard digestion in a test tube, yet stimulate the secretions of the mucous membrane or be so rapidly absorbed as to have but a passing effect in so far as they become a factor in the digestive process, or in altering either the composition or reactions of the digesting mass. In the medicinal use of the most pronounced retarding substances, they will seldom or never be so given as to impart to the digestive fluids the percentage which has been found inhibitory in the test tube. There are few soluble medicinal substances which, in some proportion, do not exhibit retarding action under test tube conditions. Experience, long in advance of these experiments in artificial digestion, has disclosed the disturbing effects upon digestion of both dietary and medicinal substances, due to conditions quite apart from those of the test tube, and in which these experiments afford but little practical significance.

INCOMPATIBLES.

SUBSTANCES AND CONDITIONS WHICH DESTROY THE DIGESTIVE FERMENTS.

It is remarkable that there exists so little difficulty in the practical use of the digestive ferments. It is not considered a hardship that nitrate of silver must be excluded from light, or anæsthetics from evaporation ; or that hypodermic solutions must be freshly prepared. Whilst the incompatibility of drugs and chemicals extends to the formation of dangerous compounds from simple mixtures, the digestive ferments have practically but one sort of incompatibility to be avoided in dispening or prescribing, that of substances or influences which render them inert. The manufacturer should not offer and the physician will not knowingly prescribe combinations, which are or are likely to become inert by the time they reach the patient's hands. Of all the conditions and substances with which the digestive ferments are brought into contact in their practical use, we may conveniently summarise those which render the ferment inert.

A digestive ferment should never be mixed with water or any fluid of a higher temperature than can readily be borne by the mouth. In the peptonising process, in "sprays," in "surgical solvents," too high temperature should be carefully avoided. Pepsin is destroyed in alkaline solutions—with lime water, sodium bicarbonate, aromatic spirits of ammonia, etc. All ferments in solution soon decompose unless in the presence of an antiseptic. Therefore, a mixture of trypsin, or pancreatic extract, water and soda can not be expected to keep indefinitely. The ferments should not be mixed undiluted with strong alcoholic tinctures, or astringents. Pancreatic ferments should not be placed in acid mixtures. Pepsin and pancreatic ferments should not be mixed together in solutions acid or alkaline. These mixed ferments can not be permanently held in an active form in any solution— elixir or whatever it may be called.

In using a digestive ferment, it does not matter whether the ferment is in solution or suspended in a mixture ; whether the substance with which it is mixed is known to retard digestion in a test tube ; the main point is that the ferment shall not be destroyed—that it be exhibited in an active form.

PEPSIN AND BISMUTH IN SOLUTION.

Bismuth in solution is incompatible with pepsin. Pepsin and the insoluble salts of bismuth, the subnitrate or the subcarbonate, is one of the most efficient and generally used combinations. Obviously these salts of bismuth exert no influence upon pepsin in the dry state, nor are they injurious to the ferment when mixed with it in the fluid form. Therefore the bismuth subnitrate may be properly given, for instance, in Fairchild's Essence of Pepsine, or in the Glycerinum Pepticum. But the soluble salt of bismuth, the ammonia citrate, cannot be combined with pepsin in solution without rendering the ferment inert, as we pointed out ten years ago. This fact has been repeatedly adduced by pharmaceutical writers, and the elixirs of pepsin and bismuth have quite lost their vogue ; there is but a very limited demand for them. Before the *digestive* valuelessness of this pepsin and bismuth elixir was known, the main attention of pharmacists was directed in the endeavor to overcome the chemical or pharmaceutical incompatibilities of this combination. This was due to the use of hydrochloric acid as a solvent, or to pepsin containing this acid, and the unstable solution of bismuth thus yielded in spite of the neutralisation with ammonia. Consequently we have advocated the employment of citric acid, which gives a satisfactory pharmaceutical preparation, and from time to time new formulas for elixir of pepsin and bismuth appear. But however combined or skillfully prepared, it will be found that the bismuth in *solution* has rendered the ferment inert. The elixirs of pepsin and bismuth are invariably found upon assay to be completely devoid of any digestive action. The good they do is from the bismuth, alcohol, the aromatics, etc. Consequently this elixir of bismuth and pepsin should be discarded, and it is to be regretted that it has found a place in any "formulas," and thus encouragement given to a palpably improper combination.

ANTISEPTICS.

The influence of antiseptics upon the digestive ferments is of great practical importance. Alcohol, glycerin and common salt are the most available technically—both in the pharmaceutical preparations of the digestive ferments and the preservation of solutions of peptonised products. Brine extracts the rennet and some of the pancreatic ferments. Other antiseptics, borax, boracic acid, salicylic acid, thymol, etc., render infusions of the ferments stable. But this class of antiseptics should not be resorted to, for everywhere in food stuffs, beverages, etc., they are distrusted and not permitted as a substitute for alcohol. They are not permissible unless directed by the physician There are antiseptics which may be so freely used in a digesting mass as to prevent all ulterior or putrefactive changes and yet not interfere with the action of the digestive ferments. Creosote is remarkable for this property and when introduced in the pancreatic digestion of milk, fibrin, etc., the usual digestive transformation takes place without the occurence of fermentative changes, even after many hours. Pure creosote is therefore justly regarded as a most valuable medicinal antiseptic to prevent fermentative changes, especially in the intestinal tract.

" JUMBLES."

There are a certain class of digestive compounds which have been aptly characterised and condemned by Fothergill as unscientific "jumbles." There is an objection to these jumbles more serious than any based upon theory as to the propriety of mixing up all the agents of digestion with acids and milk sugar. Indeed, there has never been any conclusive argument against efficient combinations of the various ferments, and many physicians employ tablets of "Fairchild's" Pepsin and Pancreatic Extract. So whatever our theory about digestive compounds, certainly a compound can be judged only by its actual digestive value, just as we judge or value the single ferments. Nothing can be easier than to triturate powders of pepsin and pancreatic ferments, and such a mixture will exhibit all the properties of each one of the ferments. Notwithstanding this, most of these compounds, "pepsin, pancreatine, diastase and acids," do not contain any other ferment besides pepsin; consequently there can be no escape from the conclusion that their value as remedies depends solely on the pepsin and acid they contain. Such compounds have been again and again condemned for their defective and deceptive character by competent medical and pharmaceutical writers. There is a fallacy that is fast being ventilated, in the pretence that such compounds possess peculiar remedial or "clinical" value, in spite of their failure to show digestive action. The use of these compounds is one of the strongest evidences of the value of pepsin, even when diluted with milk sugar.

VEGETABLE FERMENT.

The property of certain vegetable "milk juices," of softening or liquefying fibrin and albumen, has long been known. It has been of speculative interest to the botanist and physiologist ; but these exceptional instances of the presence in plants of a proteolytic ferment of no discoverable relation to their nutrition is without physiological significance. It is of practical import to the physician only to the extent in which these vegetable substances may prove to possess any superiority to the animal proteolytic ferments. These vegetable principles have been the subject of considerable experiment in various quarters without the appearance, to our knowledge, of any data showing their especial utility or availability. On the contrary, they have been deemed far weaker than the animal ferments. In order to ascertain for ourselves the properties and comparative value of the vegetable products we, years ago, obtained specimens of dried milk juice, and the so-called active principles thereof, and subjected them to assay in acidulated water, under the usual range of conditions competent for pepsin, and with alkaline and neutral water, under the conditions suitable for pancreatic extract. Subsequently we have, from time to time, tested specimens submitted to us by parties proposing to introduce the vegetable product on the market. Again recently, our attention has been brought to the subject, and we have repeated careful tests of the vegetable ferments as found in the market. As a result of these many tests, we have invariably found all specimens, papayotin, papaine, papoid, etc., of such very slight activity in comparison with either pepsin or pancreatic extract, that we have always declined to introduce the vegetable product, and have never found reason to undertake its manufacture. The vegetable ferment exhibitits no new, peculiar or superior property either as regards media or character of action. Considered as a "vegetable pepsin," its value must rest upon its action in acidulated water, for pepsin has no action except in acid ; here papayotin or papoid is practically inert. Considered as a ferment capable of action comparable with trypsin, its value must rest on its action in neutral or alkaline media ; here it is of very feeble power. As a peptoniser of beef, fibrin, egg albumen or milk, its action is so slight and unsatisfactory as to be of no practical utility. The claims advertised, that a certain vegetable product acts in acid or alkali, in "less water" are simply preposterous. Water is essential to all physiological action. The simplest tests under various ranges of acidity, alkalinity, all ranges of temperature, of proportions of water, will at once show that the " vegetable " ferment possesses no immunity from the conditions governing all ferment action. The most remarkable thing about the "vegetable pepsin " is that its value is in an inverse ratio to the claims made for it, and the prices asked for it.

THE GASTRIC FERMENTS.

PEPSIN.

Pepsin, the peculiar digestive principle of the stomach, is active only in the presence of an acid, and most potent with the acid of normal gastric juice and with the percentage of free acid present at the height of gastric digestion. Its action is, however, by no means inseparably associated with hydrochloric acid, but it acts freely with a wide range of acidity with both mineral and organic acids, lactic, tartaric, etc. Even the faintest acidity is sufficient to call forth its action, which is not greatly modified at points either slightly above or below the free acid of the normal media. Pepsin digests but one form of substance, proteids, all forms of which it is capable of converting into peptone. Various forms of proteids show some varying behavior under the influence of the artifiicial gastric juice. Coagulated egg albumen goes into solution from the surface gradually and only upon prolonged contact does the excess of albumen show any notable softening effects or gelatinous appearance. Raw fibrin or fresh meat, instantly swells in contact with acidulated water, and then undergoes softening and solution. Boiled fibrin or flesh behaves like boiled egg albumen. If, however, the raw albumen has been previously dissolved in water and then boiled, we obtain a gelatinous or mucilaginous albumen, which upon contact with the active ferment almost instantly becomes thinner and soon goes into complete solution. The ferment here behaves almost identically as does diastase with gelatinous starch, and the resulting solution will contain various forms of soluble albumen and peptone, just as the starch solution will contain soluble starch and dextrin. The peptic digestion of albumen is a gradual, progressive transformation into peptone, with various intermediate forms of soluble albumen, the percentage of peptone depending greatly upon the proportion of ferment to the albumen and the duration of the digestion. Gastric digestion ceases at peptone ; there is no further change or splitting up of the peptones as in the case of pancreatic digestion. Pepsin is not only inert in alkaline solutions, but is destroyed with merely sufficient alkali (such as for instance, sodium bicarbonate,) to give an alkaline reaction. It is not possible by subsequent acidulation or any treatment, to bring the ferment to show the slightest activity.

PEPSIN, ITS PRACTICAL USES.

Pepsin is not available for peptonising food for the sick, in the household. Its action is not only restricted to albuminous substances, but acid being indespensible, the product is for this reason unsuitable as a food. In the laboratory this ferment may be and is commonly utilised, for there the acids are separated and the products clarified, properly. But the terms " peptonised " and peptone are so fixed in the popular mind in association with pepsin, that many continue to regard a peptonised food as one made with pepsin or containing pepsin. Pepsin is useless in the artificial digestion of milk. Pepsin cannot be used for the artificial digestion of food at the table in the way that the Extractum Pancreatis may be. Pepsin is, however, useful for the solution of fibrinous membrane, coagula, etc., and is much employed as a surgical resource. For its use in this direction, see " Surgical Use of the Digestive Ferments." The exhibition of acid in conjunction with pepsin depends much on circumstances, for it cannot by any means be held to be always indispensable. Normal gastric juice contains both free acid and pepsin. An artificial gastric juice for digestion in a flask, can only be obtained by combining these two agents. But the stomach cannot be dealt with as with a beaker glass. We see the good effects from the administration of pepsin without acid or in such faintly acid solutions as Fairchild's Essence of Pepsine. If a physician sees indication for the administration of soda, as in acidity of the stomach, and for pepsin to aid digestion, these two remedies may be combined dry without regard to the fact that in an alkaline solution pepsin is inert. For it is not supposable that an amount of soda sufficient to impart an alkaline reaction to the entire gastric contents would ever be given.

MILK CURDLING FERMENT OF THE GASTRIC JUICE.

The gastric juice contains a distinct principle which has the power of curdling or coagulating milk. It is not known

to have any other property ; consequently it cannot be considered a digestive ferment in the sense that it effects any
change in the constitution of an alimentary substance.
Whilst many studies have been made and theories advanced
concerning the action of this ferment, of the changes milk
undergoes in coagulation by it, we can only say of this
ferment as of others, we do not know how it acts ; we know
it to be a true ferment. A solution of·this ferment heated
to 170° instantly loses all activity. Its action is not, like that
of pepsin ferment, dependent upon the intervention of an
acid. It curdles neutral or even faintly alkaline milk. It may,
like pepsin, be extracted by acidulated water. Therefore, we
may obtain both ferments in an active, permanent solution.
In remarkable contrast to pepsin it is not precipitated by
common salt. A brine extract is commonly employed in
the use of " rennet " in cheese making. These salt infusions
of the rennet are devoid of all peptic activity. Pepsin has
no curdling property, and whatever milk curdling action a
preparation of pepsin may show is due entirely to the
true curdling ferment associated with it. Notwithstanding this fact, the impression still obtains that pepsin should
curdle milk. The question has been raised as to the value
or function of this ferment in the natural process of digestion, seeing that the gastric juice contains acid which
itself, it is said, should coagulate milk. But it is a fact that
milk does not by any means behave with acid precisely as
with the curdling ferment ; in the one it is entirely a chemical change, in the other a physiological change wrought
instantly, and accompanied by no change in the chemical
constitution of the caseine. Further it is of great significance, it seems to us, to find that this ferment exists in the
greatest activity in the stomach of the suckling animal.
This is so well known, that it is always the " milk rennet "
which is used from which to prepare rennet liquids. The
stomach of the hog contains but a trace of the ferment, and

pure pepsin from this source is invariably useless for curdling milk. It would, therefore, seem quite likely that this ferment plays no insignificant part in the digestion of milk. By its action the caseine is thrown out in the form of coagula, most susceptible to the action of the gastric juice, whilst the whey, containing the salts and milk sugar and the soluble forms of albuminoid, passes freely along the digestive tract, where it undergoes assimilation without the need of any digestive action. In the young suckling, furthermore, the pancreas is but partially developed. In this curdled milk, we see that the caseine is reduced to a condition analogous to that in which the flesh foods of the adult are presented to the stomach. If the milk were not curdled, either by acid or rennet, certainly there would be no obstacle to the free passage of the fluid milk along the infant's digestive tract. It is not, then, without purpose, this curdling ferment in the stomach of the suckling animal.

The liquid rennets of the shops prepared from salted vinous menstruum do not contain pepsin and have been so inferior and variable in curdling activity that they have fallen into disuse.

THE PANCREATIC FERMENTS.

TRYPSIN.

THE PROTEOLYTIC FERMENT OF THE PANCREAS.

The pancreas juice contains a principle which may be said to be the analogue of pepsin, in that it is capable of converting all forms of proteids into peptone. It differs markedly from pepsin in important particulars. Whilst it is most active in an alkaline solution, it is also energetic in a neutral solution, and digests milk freely without addition of alkali. Thus, it is not restricted in its media to the reaction characteristic of the fresh pancreas juice, having on the contrary, as will be seen, a wide range of action. Whilst in a feebly acid solution, especially with organic acids, it is found to exhibit action upon fibrin and albumen, free hydrochloric acid is far from being favorable to tryptic action.

Trypsin yields various soluble products and peptones which do not materially differ from those of peptic digestion, but unlike pepsin it gives still further normal products by the transformation of peptone into leucin and tyrosin.

Trypsin has a special affinity for the proteids of milk, showing proportionately more activity upon caseine than upon other proteids. In the Extractum Pancreatis, trypsin is presented as naturally associated with the other ferments of the gland. We have also made a special preparation of trypsin as a solvent for diptheritic membrane.

THE DIASTASE OF THE PANCREAS.

The starch digesting principle of the pancreatic juice presents no known difference from the ferment of the saliva, or of germinating grain (malt) in the media or method of its action, or in the result of its action. Like every known form of diastase, it gradually converts gelatinous starch into soluble starch, dextrines, glucose or malt sugar. It is active in neutral or alkaline reactions. We cannot therefore, distinguish one form of diastase from another merely by its behavior, and whatever views may be held regarding diastase as a remedy, must apply to this ferment from whatever source it appears. We have no means of knowing definitely to what degree of conversion starch is carried in the natural process of digestion, but there is no doubt that its complete transformation into glucose is not essential to its assimilation ; physiological experiments show that the highly diffusible dextrines and soluble starches are absorbed into the blood, and there is every ground to suppose that a very considerable portion of the products of starch digestion are absorbed long before they could reach their ultimate conversion. The influences of various substances upon diastase are, as in the case of all ferments, largely modified by the proportion of ferment and by the presence of products of digestion, etc., so that, whilst in the laboratory we fix the retarding influence of definite percentages of acids or of alkali under arbitrary conditions, we must not overlook the insufficiency of such data as bearing upon the actual conditions of digestion within the body.

The incompatibility of acid and diastase does not afford sufficient ground for the assumption that the starch digesting principle finds no field for action in the stomach. The gastric juice is absolutely inert upon starch ; the major part of farinaceous matter is in a form incapable of solution during the short contact with the salivary diastase in the mouth ; the stomach contents have at the outset of normal digestion but a feebly acid reaction, and the acidity only reaches its maximum point an hour or so after the ingestion of food. In view of all these facts, the conclusion is reasonable, that the stomach affords opportunity for such preliminary digestion of starch as fits it for further conversion in the intestinal tract.

USES OF PANCREATIC DIASTASE.

The practical identity of the pancreatic and salivary diastase being established, it follows that we may as reasonably exhibit an active extract of the pancreas in deficiency of salivary digestion, as we may exhibit pepsin in feeble gastric digestion.

The most rational way of supplementing deficient salivary digestion is to add the active pancreatic preparation to farinaceous food at the table. No taste is imparted to the food, no suggestion whatever of "medicine," and under its influence the starch is rapidly softened, and converted into a soluble form which will insure its proper digestion. For children and the aged and convalescent, this method is especially recommended. For faulty intestinal digestion of starch, the Extractum Pancreatis or Diastasic Essence should be given immediately after meals, and repeated in an hour or two. In the Extractum Pancreatis the starch digesting principle is accompanied with ferments which digest all other forms of aliment, and which are often indicated in connection with diastase. In the Diastasic Essence of Pancreas, the starch digesting principle is presented in an exceedingly active, agreeable and practically isolated form, and it, therefore, may be advantageously resorted to in cases where it is not desired to adminster the other ferments of digestion. For intestinal indigestion of starch, the diastasic ferment may best be given just previous to taking food, and again about two hours after food. The diastasic ferment given just previous to or with meals promotes the preliminary starch digestion, that which is normally effected by the salivary diastase ; given after the force of gastric digestion is lessened, it promotes the secondary or pancreatic digestion of starch. The Pancreatic Tablets and the Diastasic Essence are especially commended as a means of exhibiting this ferment.

THE EMULSIVE FERMENT.

The characteristic action of the emulsive ferment is the conversion of oils or fats into a minute state of division or

emulsification. The emulsification and absorption of the
minute divided molecules are successive steps, precisely as
the conversion or hydration of albumen is antecedent to its
absorption. Whether there is also any chemical change in
the fat by action of the ferment, may be said to be a moot
question. If a fat or oil is macerated for some hours at
the temperature of the body with fresh pancreas juice or
with minced pancreas and then strained, it will be found
that the pancreatised fat will instantly form a thick, creamy
emulsion when shaken with an equal quantity of water.
The pancreatic juice when taken from the gland soon undergoes change and shows an acid reaction, and it has been
asserted that when the development of these fatty acids is
prevented, there is no occurence of these acids in the
treatment of fat under the influence of the emulsive ferment. As the chief and characteristic behavior is the
breaking up of the fat into emulsion or creamy form,
so doubtless is the greater part of fat assimilated
direct without undergoing any conversion. It is a
curious fact that this, the least available and the least
practically important pancreas ferment, has been the
one to which attention had been chiefly directed prior
to the introduction of the Extractum Pancreatis.
The "pancreatines" found in commerce bore no
other reference to any digestive action or use than
their asserted property of emulsifying cod liver oil,
whilst they were completely deficient in the other ferments of the gland. The emulsive ferment is not
capable of effecting the permanent admixture of oil
with water, as may be done by purely mechanical
agents, such as gums, etc. The pancreatic liquors,
as well as the commercial pancreatine, are of only
the slightest emulsive value. The Extractum Pancreatis is in this, as in other ferments of the gland,

the most active product. If a few grains of the Extractum Pancreatis be well shaken with one or two drachms of warm water, and an ounce of cod liver oil or pure olive oil, and allowed to stand in a warm place for five or six hours, it will be found that this oil will then form a creamy emulsion with water; upon standing, this emulsion will gradually separate, but may again be emulsified by agitation. When the use of the gummy and starchy, sweet emulsions are contra-indicated, The Extractum Pancreatis may be utilised as already described, or it may be given immediately after the pure oil or the usual emulsions.

THE MILK CURDLING FERMENT OF THE PANCREAS.

The pancreas contains a ferment which curdles milk slightly acid, neutral or alkaline. As associated with the other ferments of the pancreatic juice or in an active extract of the pancreas, such as Fairchild's Extractum Pancreatis, it cannot be practically utilised in the same manner as the rennet ferment in the preparation of curds and whey.

The proteolytic ferment will attack the curded caseine and soon dissolve it. If a few (five) grains of Extractum Pancreatis be added to pure lukewarm milk, (say four ounces) a soft curd will be almost instantly formed. If the milk is permitted to stand at rest, the curd will not cohere and separate in a mass from the whey as in rennet curdling, but will gradually become softer, will float in the milk and finally disappear. If the milk is stirred with a rod or spoon, the curd is instantly broken up into minute particles, blended with the milk and soon undergoes digestion, the milk acquiring the characteristic color and taste of peptonisation.

THE DOSAGE OF DIGESTIVE FERMENTS.

The digestive ferments having no drug action, no property comparable to that by which the doses of remedial agents are in general regulated, a small dose differs from a large dose only in degree, not in character of effect produced. There is no such relation of effect to dose as in the case of drugs, such as ipecac, calomel, strychnine, etc. We cannot expect, therefore, to fix any arbitrary range of dose as with drugs with distinct measurable action upon the body. In the days of the saccharated pepsins and pancreatines, large bulk with minute quantities of true ferment were given ; with the introduction of Fairchild's preparations of unprecedented activity, a few grains became the generally employed dose, and the tendency seems to be, as these products are more and more improved in potency, that the doses are rather diminished. Doubtless there has obtained in the past some impression that large doses of pure ferments might not be harmless ; but neither in medical literature nor in anything that we know of the physiology of digestion, nor from the extended opportunity for learning the results of the practical use of the digestive ferments, does there appear any tenable ground for this assumption. It has existed only as a vague theory, and as a surmise of possibilities. The animal digestive ferments find a place in materia medica, because they display upon food substances under conditions closely conformable in temperature and in reaction to those of the body, the action characteristic of the normal digestive secretions. Properly introduced into the living digestive tract, we may then expect that they will exert precisely the same effects as the naturally secreted ferment. A large dose of pepsin artificially introduced into the process of digestion can no more attack the stomach membrane than will the natural gastric juice. The sufficient dose to supplement deficient digestion must vary largely and the dose need only be regulated by considerations of the amount required to effect the purpose. That habitual use of a ferment may rationally be resorted to, in order to produce tranquil digestion for those patients whose digestion is susceptible to disorder, or impaired by care, anxiety and sedentary occupation and similiar influnces is beyond question. What expedient more practical or innocent ? Certainly far less likely to be harmful, than persistent " drugging."

THE FAIRCHILD PREPARATIONS

OF THE

DIGESTIVE FERMENTS.

The uniform character, activity and reputation, of the Fairchild Preparations of Digestive Ferments are sufficient evidence, of the fallacy of the statements sometimes advanced, that the digestive ferments are necessarily variable and unreliable agents. The fact is that Fairchild's digestive ferments are second only in uniformity to the alkaloids and chemicals. They are more definite and uniformly reliable than most drugs, or galenical preparations therefrom,—extracts, tinctures, etc. The status of these preparations of the digestive ferments, moreover, does not depend simply upon medicinal properties, so difficult to determine for all agents except those which have a distinct action upon the body. They are valued for reason of definite demonstrated and applied digestive properties.

The Fairchild special products have been put forward with definite methods for accomplishing certain practical results. As a record of more than ten years of experience, not an instance has occured where one of our preparations has failed to perform this work—has disappointed the anticipation of the physician. Not a package of the Extractum Pancreatis nor a Peptonising Tube has proven inert upon milk or starch. The surgeon who sees the potent solvent action of Fairchild's pepsin upon morbid tissue ; the physician who sees the absolute certainty with which the Extractum Pancreatis may be applied in peptonising milk, etc., will not doubt these agents possess substantial claims to therapeutic use.

The Fairchild preparations are the result of original special work given to the digestive ferments. Each and every product offered to the medical profession has been carefully prepared to meet certain requirements and to fulfill a definite purpose.

We have especially sought to avoid all incompatible compounds, and do not supply these even if having popular sale, and do not needlessly multiply the variety of preparations. For all the important purposes for which the digestive ferments are now applied, the Fairchild preparations have been either originated, or been the means employed, on account of their well-known superiority.

The commercial products will differ in character, in grade of activity, in purity according to the skill, knowledge and purpose of the manufacturer. Some are made to supply a demand for cheap preparations ; others to imitate in physical characteristics products which have become eminent for value.

The Fairchild preparations being almost exclusively dispensed upon the prescription of the physicians, we have with a due regard to the interest and convenience of the pharmacist, supplied them in bulk whenever practicable. This fact has, however, only afforded better opportunities to that class of manufacturers who find it impossible to make a market for their goods on the score of merit, but who, as in every " line," prefer to trade upon the reputation of standard products by the substitution of inferior and "cheap" imitations. These inferior products are urged upon the pharmacists as the " same thing," " gives you extra profit ;" if " Fairchild's is not specified, use ours ;" although it is plain that the physician even when not specifying Fairchild's really means and expects to get Fairchild's, owing to his long reliance upon them years before the appearance of so many preparations under the same titles.

Owing to the great reputation and use of Fairchild's Essence of Pepsine, this preparation has been the especial object of imitation.

The title, Essence of Pepsin, has been applied to preparations entirely dissimilar and inferior in properties to the original Essence of Pepsine.

In many instances, upon complaint of physicians, we have examined these dishonest imitations which have been substituted even when Fairchild's is specified, and found them often inert upon milk, of weak peptic power and of a distinctly unpalatable character. In view of the very important properties and uses of this Essence as a means of administering drugs, preparing whey, etc., to correct digestive disorders of infancy, the substitution of these worse than useless preparations inflicts serious injury—injury upon the patient, the physician, and the manufacturer upon whom the physician relies. Further the *same price* is charged the customer as for the original article—which leaves no question as to the real purpose of this infamous practice. We desire to take this occasion to say that in contrast to this, the great body of pharmacists not only religiously regard the wishes of the prescriber, but many of them, we are glad to know dispense, and use generally, Fairchild's preparations by their own preference. Seeking no other market than that caused by the preference for Fairchild's preparations, we desire to protect ourselves, the physician and his patient, from the substitutes and imitations. We, therefore, ask the favor, that the physician will distinctly specify Fairchild's when he wishes them, and in case of any dissatisfaction that he will send to us for examination the preparation dispensed.

FAIRCHILD'S

PEPSIN IN SCALES AND POWDER.

Pepsin in Scales, "free from all added substances," was originated by Fairchild Bros. This, the first pepsin in scales, was vastly in advance of the commercial products then in the market, and for some years it was the only pepsin known in the *form* of scales and by this *title*. But the reputation and use it obtained ultimately invited the inevitable imitations of name and appearance. The poor quality of many of these made it plain that they were made to supply the demand for pepsin in scales. That the reputation and value which became attached to the product and the name "pepsin in scales," are due to the quality of the original Fairchild's Pepsin in Scales, is beyond question. It has always possessed the two essential qualities of dry pepsin—activity and permanency. Fairchild's pepsin is just as permanent in powder as in scales. The powder is simply the powdered scales. The powdered peptone products are very susceptible to moisture. Fairchild's pepsin is especially suitable for dispensing in powders, capsules, etc. It will not become sticky, even in damp climates—will not gum up in the powder papers, either pure or mixed with bismuth, soda, etc. One grain of Fairchild's pepsin will digest 2,500 grains egg albumen in six hours at 105° F.

SACCHARATED PEPSIN.

This, the only officinal form of dry pepsin; is convenient for those who wish to prescribe pepsin in very small doses, as for infants. But its digestive standard is so weak that there appears little use for a saccharated product, in any case where it is at all practicable to mix the pepsin and milk sugar in whatever proportion desired by the prescriber. The officinal saccharated pepsin represents about two and one-half per cent. of 1 to 2,000 pepsin, about 98 per cent. milk sugar, thus but a minute quantity of actual ferment in a dose of practicable bulk. Further, when a physician prescribes pepsin without specifying which product desired, the druggist must try to decide whether "pure" or saccharated is required, or simply dispense the officinal. Most of the "saccharated" sold does not even conform to the U. S. P. standard, which requires 1 grain to digest 50 grains egg albumen.

GLYCERINUM PEPTICUM.

(FAIRCHILD.)

Glycerin possesses peculiar value as at once an extractive and preservative of the digestive ferments. For this purpose it has long been used in the physiological laboratory and the glycerin extracts have been preferred for experimental purposes. Fairchild's Glycerinum Pepticum is the first commercial product in which glycerin has been utilised to prepare a concentrated, stable solution of pepsin, direct from the mucous membrane. This Glycerinum Pepticum presents the peptic ferment in the most isolated form in which it has ever been produced in solution for practical purposes, containing no alcohol, sugar, flavoring or antiseptic other than the pure glycerin. It is a clear, bright extract, remarkably free from color, odor or taste, freely soluble without cloudiness in all proper menstrua or media. It is notably devoid of the peculiar disagreeable characteristics of the glyceroles of peptone pepsin. It is by far the most convenient and useful for all purposes where pure pepsin is required in solution ; for extemporaneous mixtures, for experimental purposes, for applying pepsin as a surgical solvent, for preparing officinal solutions. It is quite agreeable, even in pure or acidulated water and may be given in wines, elixirs, etc. It is especially convenient for the physician who finds it desirable to dispense medicines and for hospitals and dispensaries. For the manufacture of all the usual pepsin fluids, wines, elixirs, liquors, etc., it is far preferable, gives a more stable and agreeable preparation than obtainable by any other form of soluble pepsin. Twelve minims are capable of digesting 2,000 grains albumen under usual conditions.

In Fairchild's pepsin in scales or powder and in Fairchild's Glycerinum Pepticum, the physician and the pharmacist have the dry and soluble ferment in the most convenient and desirable forms, covering all possible uses of the concentrated pepsin. The peptone-pepsins have never filled both purposes of perfect solubility and resistance to moisture.

ESSENCE OF PEPSINE.

(FAIRCHILD.)

A SOLUTION OF THE ESSENTIAL ORGANIC INGREDIENTS OF THE GASTRIC JUICE, EXTRACTED DIRECTLY FROM THE PEPTIC GLANDS OF THE STOMACH.

Fairchild's Essence of Pepsine is obtained by direct extraction from the fresh calf rennet in a menstruum which possesses, in the highest degree, the properties of a vehicle and a preservative of the peptic and the milk curdling ferment.

The Essence of Pepsine is a remarkably agreeable, diffusible, aromatic stimulant ; yet holds in solution both the active ferments of the fresh gastric juice. It is but faintly acid, not in the least heavy with sweet, leaves upon the palate not the least pronounced impression of any pre-dominant flavoring. It is free from all suggestion of animal origin, and further, imparts a delicate flavor and aroma to milk-curd or junket, whey and cold milk.

Pepsin, like many another remedy, gains by judicious association of corrigents, and only second to the actual agents of digestion are the aromatics skillfully combined. It is a matter of common experience, even in health as to the influence of savory substances, and of the remarkably malevolent effect of some discordant flavor ; and dyspeptics are especially sensitive to digestive disturbances out of all proportion to the apparent cause. The usual pharma-ceutical preparations, the elixirs, wines, cordials, and so forth, are generally poor examples of what a blend of proper aromatics should be. Many of these preparations, produced in imitation of Fairchild's Essence, are so dis-tasteful as to greatly militate against the good effects which might be derived from any ferment they may contain.

Fairchild's Essence of Pepsine has long been the most

useful and successful of all pepsin preparations. It is found of peculiar value for three distinct purposes—as a remedy for indigestion in adults and infants ; as a means of administering drugs which disturb the digestive functions and impair the appetite ; as a practical rennet agent.

For infantile digestive disorders Fairchild's Essence of Pepsine is especially effective ; it not only aids digestion, but corrects flatulency and vomiting. It is, therefore, far more innocent and effective than the usual domestic and empirical remedies for colic, etc. ; certainly infinitely preferable to soothing cordials. In cholera infantum it presents the most valuable properties—stimulant, carminative, and digestive, and is far better than alcoholic stimulants, per se. For persons of habitually weak digestion it proves the most acceptable and potent resource.

The usual dose for an infant is from 5 to 10 drops, and from 1 to 3 teaspoonfuls for an adult.

That Essence of Pepsine is of great service in aiding the tolerance of drugs such as iodides, bromides, mercurials, etc., is well known. Here it is not only important to give pepsin, but the ferment must be in such a form as to overcome the repulsion caused by the ingestion of these drugs. The Essence of Pepsine has proven of the greatest possible service in the administration of such drugs, because of its digestive and grateful stomachic properties. It is confidently relied upon for this purpose by many physicians. As a vehicle simply, the Essence is by far the best in use. Drugs which give unsightly mixtures or which completely overcome the agreeable qualities of the Essence should be given in separate form, to be immediately followed by one or two teaspoonfuls of the pure Essence, thus gaining the greatest advantage of its agreeable qualities.

The efficacy of the Fairchild's Essence for administering

iodide of potash and the certainty with which the Essence is used in making milk curd or junket, has led to the successful experiment of using this essence junket itself as a vehicle for the iodide of potash. The method of preparing this medicated junket is given on page 45.

Further practical uses of Essence of Pepsine in preparing whey, junket, etc., as food for invalids and in cholera infantum, are given in *Practical Recipes*.

Fairchild's Essence of Pepsine was the first medicinal preparation ever offered of the two gastric ferments, pepsin and milk curdling. Such a pharmaceutical product from the fresh stomachs has only been obtained by many years of experience and skill, and utmost care and nicety in manipulation.

The wine and elixir of pepsin obtained as they have been, by dissolving absurdly small proportions of saccharated or other pig pepsins in wine, etc., are practically useless. The more recent class of " essences," etc., made in imitation of Fairchild's, to fill the prescriptions for Essence where Fairchild's is not actually specified, are greatly inferior in every respect to the original Essence of Pepsine. They are obviously made from peptone pepsins dissolved in " elixir bodies," are inferior in every important respect, quality, flavor, pepsin and rennet action. Many physicians who have for ten years and more used Fairchild's Essence, naturally expect this will be dispensed when they order Essence of Pepsine, but it is now very important to specify Fairchild's. The substitutes cost the patient the same price as the genuine.

MEDICATED JUNKET.

JUNKET WITH POTASSIUM IODIDE, MERCURIALS, ETC.

The use of junket as a vehicle for the exhibition of iodide of potash was first suggested by Dr. D. Bryson

Delavan of New York, in a paper which appeared in the *New York Medical Record*, Nov. 28th, 1891. Dr. Delavan found by dissolving the iodide in Fairchild's Essence of Pepsine, and adding this to a small quantity of warm milk, that the curd which was instantly formed completely enveloped the salt in a thoroughly diffused form, and gave no taste or suggestion of its presence. It was found by this writer that the iodide could thus be freely administered without the disturbance of digestion so characteristic of this most distinctly repulsive chemical.

Subsequently we tried Fairchild's Essence of Pepsine, used in the manner suggested, in preparing junket with Potassium Bromide, Sodium Salicylate, Iodide of Potash with Biniodide Mercury, Chloral Hydrate, etc., and with all these found the Essence to at once yield an agreeable, jelly-like curd. We have in this junket, as prepared with Fairchild's Essence of Pepsine, a distinct acquisition to our means of administering a class of drugs which it is of the utmost importance to be able to give in a form which does not disturb the stomach. Wherever it is desirable, this junket may be used to convey the iodide, mercurials, etc., without the knowledge of the patient as to its medicinal character. This is the formula best adapted for prescription :

> Potassium Iodide....................f ℨ ii.
> Essence of Pepsine, Fairchild's.......f ℥ iii.

Add one teaspoonful to a wine-glass of warm milk, and take the resulting curd (after meals, or at such times as desired to order it). From 5 to 10 grains Salicylate Soda, or Potassium Iodide, with $\frac{1}{10}$ to $\frac{1}{8}$ grain Mercury Biniodide, may be ordered to each fluid drachm of the Fairchild's Essence.

PEPSIN TESTING.

Notwithstanding the immense study which has been given to pepsin, no satisfactory chemical test for it has

ever been established. The more we learn about the
digestive ferments, the stronger becomes the conclusion
that they are all some form of albuminoid matter, as they
are themselves the product of albuminoid cells. A chemical test for pepsin must be one very sharply distinguished
from all other reaction of albuminoids; must be the unmis-
takable evidence of the living ferment. For a ferment
may have been subjected to influences which may have
quite destroyed its activity and not appreciably altered its
physical or chemical characteristics. Whilst we may discover some peculiar reaction for pepsin, it is scarcely
possible that we can ever assay pepsin by chemical analysis.
As pepsin appears in commerce, the ferment is associated
with substances readily distinguished, such as common salt,
milk sugar, hydrochloric acid and starch. If these are in
obviously large proportion, the inference will be that the
products are weak, yet this is by no means certain, for a
saccharated pepsin may prove more active than a so-called
"pure pepsin," in which the ferment is either injured in the
process of manufacturing, or presented in a very large
proportion of gelatin, albumen or peptones. Gelatin,
albumen, etc., have been employed in the manufacture of "pure scale pepsin;" the main object being to
trade upon the reputation of the original Fairchild
pepsin in scales the value and reputation of which
have made the title scale pepsin of such commercial
importance. The peptone is the product of the self-
digestion of the lining membrane in acidulated water with
heat; the ferment thus dissolves the proteid matter in
which it is secreted, or which may be added, just as diastase
dissolves the starch in the germinated barley or malt.
There is thus a distinct analogy between maltose (malt
extract) containing free diastase, and peptone containing
free pepsin. The peptone is objectionable to just that
degree that it dilutes the pepsin, renders it hygroscopic and

prone to spoil. The chemical treatment, the condensation by heat of the peptone solution to a scaling consistence may account for the great variation in activity of some of these products possessing practically identical physical properties. At present we can only test pepsin by its action on albuminous matter in acidulated water. The form of albumen most uniform, convenient and satisfactory is the white of egg. It is but a few minutes' work to form some opinion of any brand or specimen of pepsin by ascertaining if it has any marked action on gelatinous egg albumen in warm acidulated water. Gelatinous albumen, made by dissolving fresh white of egg in cold water and boiling well and adding the HCl. digests much more rapidly than coagulated albumen. It forms a thick, opaque mucilage, very similar to gelatinous starch and behaves with pepsin just as gelatinous starch does with diastase ; the active ferment converting it instantly into a thin, watery solution. Thus the physician or pharmacist can at least readily discover an inert or worthless product. To determine the actual digestive power of any product, it is necessary to test it upon albumen under definite conditions well known to be favorable to the action of the ferment, employing a sufficient quantity of albumen to leave such an excess as to make sure that the ferment has exhausted its activity. There is a very important relation between the proportion of acid, water and albumen. This should be so adjusted as to give a definite, proper percentage of free acid in the mixture. The albumen alkali neutralises a certain amount of the acid ; the acid forms certain combinations with the albumen at various stages of digestion. A test mixture may appear to have less acid than another, yet have more acid, owing to the small proportion of albumen to acidulated water and *vice versa*. The parts of acid to parts of water must then be regulated according to the proportion of albumen to acidulated water. The U. S. P. test, for instance, has less than the usual albumen and a greater proportion of acid to a given volume of water, and has thus too much free acid and is not a well adjusted test. In tests so adjusted as to start with exactly parallel quantities of ferment and albumen and acid, the results will vary with the volume of water. The larger the proportion of water within limits, the more digestion ; with too little water the fluid soon becomes clogged with the products of digestion. The acid hydrochloric U. S. P. of commerce varies in percentage of absolute acid. Therefore it is well to employ acid of a known strength and to use the same specimen of acid in a series of experiments. The quantity of albumen necessary for a series of tests should all be prepared at one time. The water, acid and albumen always mixed before adding the pepsin. Pepsin acts upon albumen at from 60° F. to 140° F., and the rapidity of digestion keeps pace with the temperature up to 130° F. At blood heat the results give more significance as to the effect of the ferment in the body. Five or six hours, the usual time, also provides ample opportunity for a full practical test. But the pepsin which gives best results at 105°, will give the best results at 130° and it would be practicable to make a test which at 130° might be equivalent to the usual test, be much quicker and equally reliable.

It is sometimes proposed to test pepsin by the amount of peptone formed. This, whilst theoretically exact, is impracticable and unnecessary. There are a variety of soluble derivatives of albumen not well understood and difficult and tedious of assay even with experience. Solution is the characteristic effect—the pepsin which converts the most albumen into solution is the most active and will have formed necessarily the most of all soluble products, peptones, etc. No two published tests at the present time call for exactly the same proportions ; the same product will give varying results in each test. It is very important, therefore, that we should have a standard test and that the digestive power stated for each and every pepsin, or preparation of pepsin, should be that determined under the standard conditions. Whatever the quantity of albumen used (according to the strength of the pepsin) the ratio, the proportion of albumen, water and acid should always be the same. We have employed these proportions : Coagulated Egg Albumen, 150 grains (10 Gm.), Water, 1 fluid ounce (29.7 c.c.), Hydrochloric Acid, 5 minims (0.3 c.c.)

In the comparative tests it is essential that the conditions shall be exactly alike in each test. The mixtures should all be prepared cold in bottles of the same size and well shaken to secure uniform conditions before adding the pepsin. Then the ferment added and all the flasks placed at once in a warm chamber with constant temperature of 105° F. on an automatic shaker. The pepsin should be weighed and added direct to each flask. It is impossible to reach accurate results in comparative tests by triturating the pepsins with water and taking definite amounts of the solutions or mixtures. The products vary so much in solubility that the more freely soluble, by this method, are at an advantage. The proportion of albumen dissolved, the percentage left at close of test affords fair evidence of the digestive value of each specimen.

EXTRACTUM PANCREATIS.

(FAIRCHILD.)

This extract of the pancreas presents all the active principles of the gland in the form of a dry, whitish powder. It is not an artificial compound, it is absolutely free from all added substances, and contains the ferments as they are naturally associated.

The Extractum Pancreatis which we originated in 1881 was the first product offered to the medical profession containing all the pancreas principles in a pure, dry powder. It was originally far more active and available than any other pancreatic product (the pancreatines were practically useless), and since then, the Extractum Pancreatis has shown the result of the persistent efforts to increase its efficiency and refinement. It is not too much to say that probably no remedial agent introduced during this decade has been of greater importance and value in practical medicine. Whilst the complex digestive action of the pancreatic juice was well known to physiologists, it had been but little utilised previous to the introduction of Fairchild's Extractum Pancreatis.

By means of the Extractum Pancreatis, the pancreas ferments are now effectively administered and are steadily advancing in repute as therapeutic agents. The Extractum Pancreatis has further been the means and tne basis of all the great progress made in the peptonising process, which has revolutionised the feeding of the sick and provided the long sought means for the conversion of caseine to the requirements of the infant's stomach and to the standard of mothers' milk.

As a solvent for diphtheritic membrane and as a "surgical solvent," the Extractum Pancreatis has been so successfully applied, as to merit far more extended use and promise a still wider increase of utility. The Extractum Pancreatis is by far the best simple product from the gland and inasmuch as "pancreatines" are often unfit for medicinal uses and are for the most part valueless, it is well worth while to avoid disappointment by specifying Extractum Pancreatis, Fairchild.

The Extractum Pancreatis presents all the digestive

ferments of the pancreas in an exceedingly active form
—viz.:

TRYPSIN, *which converts albumens* (*of Milk, Beef, Fish,
Blood, etc.*) *into Peptone in either neutral, alkaline, or
slightly acid media.*

DIASTASE, *which converts starches into dextrines and
sugar.*

THE EMULSIVE FERMENT, *essential to the assimilation of fats and oils.*

THE MILK-CURDLING FERMENT.

*This EXTRACT OF THE PANCREAS contains all
these digestive principles in such a degree of activity that their
presence and their action upon various food substances can be
quickly demonstrated.*

EXTRACTUM PANCREATIS.

AS A REMEDY PER SE.

"The pancreatic secretion is the most energetic and general in its
action of all the digestive juices. It unites in itself the action of the
saliva and the gastric juices, besides having properties of its own."—
T. LAUDER BRUNTON.

In view of the very partial transformation of carbo-
hydrates and proteids by the salivary and gastric juices,
preliminary to further and complete digestion by the
pancreas juice, it would seem that an active extract of
pancreas should possess remedial value of corresponding
importance. The use of the pancreas ferments as aids to
digestion has, however, been much prejudiced by theoretical views, and especially by the erroneous impression that
they are only active in alkaline media. The questions as
to whether the pancreatic ferments are capable of exerting
any influence upon food in the presence of the gastric
juice, the effects of the gastric juice upon them, have

been the subject of much experiment and discussion, resulting in conflicting theories and conclusions ; some asserting
that the pancreas ferments can resist the gastric juice,
others that they are therein rendered permanently inert.
That in the flask the pancreatic ferments are destroyed by
free hydrochloric acid plus pepsin, is, we think, beyond
question. In discussing the compatibility and value of
solutions of the mixed ferments (gastric and pancreatic) we
pointed out the fact that in a solution with pepsin and acid,
the pancreatic ferments gradually become inert, the practical
lesson being plainly that the chemist should not offer, nor the
physicians accept, remedies of this class. But to determine
the bearing of these facts upon the exhibition of the pancreatic ferments it is necessary to consider (as in the case
of all such experiments) the relations which the test tube
conditions bear not only to those of normal digestion,
but to the abnormal conditions which call for the administration of the ferments. Our present knowledge
of the phenomena of normal gastric digestion plainly
shows that there are opportune intervals for the presumably
effective introduction of the pancreatic ferments. This is
admitted even in the most conservative estimates of their
utility. There is the resource of specially adapted
pharmaceutical products. There is a marked difference
in the nature and degree of the acidity of gastric juice
during stomach digestion. There are always organic
acids set free from the food, and thus replacing a portion of the hydrochloric acid ; the hydrochloric acid is
not free and uncombined, and the gastric juice does
not correspond in its behavior to a simple solution of
similar percentage of free hydrochloric acid in water.
The presence of the products of digestion both proteids and carbo-hydrates may greatly influence the behavior of these ferments when brought into contact. In
a word, in the phenomena of digestion we have factors

which materially differ from those of laboratory experiments and must necessarily, therefore, qualify the deductions therefrom. Among those physicians who have given practical trial to the Extractum Pancreatis in intestinal indigestion, carefully regulating the mode of administration, there exists no question of its distinct therapeutic value. The feebler the digestion the less the question of interference of the gastric juice is to be considered. The Extractum Pancreatis is to be regarded first, as a diastasic agent ; second, as a digestive of albuminous food ; third, as the only means of administering the ferment which digests fat. Foster's Physiology says, " there is no means of distinguishing the amylolytic ferment of the pancreas from ptyalin." Therefore, the Extractum Pancreatis may be given as aid to digestion of starches— either at the outset of, or at proper intervals after gastric digestion. Given at the interval after eating, when the gastric action has subsided, and the ingesta freely passing into the duodenum, the pancreatic extract (in the form of tablets preferably) may be effectively administered in intestinal indigestion. In cases of almost complete abeyance of the digestive functions, as in fevers, etc., the stomach affords the necessary conditions for the action of the pancreas ferment which may be given mixed with a suitable food, such as milk, cold or warm. In such cases, it will be found on trial that no preliminary digestion (peptonisation) is necessary to insure the proper conversion of the food without taxing or disturbing the stomach itself. This is not stated on the basis of theory, but as supported by actual clinical experience.

The Extractum Pancreatis may be given in three to five grain doses—in powder mixed with food, in capsules, or Fairchild tablets, or in suitable combination. The Extractum Pancreatis has been, upon theoretical and practi-

cal grounds, recommended in the treatment of diabetes.
As a rational remedy for insufficient salivary digestion, for
intestinal indigestion, it is constantly gaining the confidence of the profession.

TRYPSIN.

(FAIRCHILD.)

ESPECIALLY PREPARED AS A SOLVENT FOR DIPHTHERITIC MEMBRANE.

This product presents the proteolytic ferment of the
pancreas in the most active form obtainable.

Trypsin has the property of digesting fibrin with
great rapidity.

It is most effective in a slightly alkaline solution, but
may be effectively applied direct to fibrinous membrane,
etc., either dry or in pure water.

It is an entirely innocent and non-irritant substance,
and does not attack the healthy or non-fibrinous tissue.

In its application to the throat all the conditions are
favorable to its physiological action.

Trypsin will be found to be a powerful solvent of
diphtheritic membrane in all cases in which it is practicable to bring it in contact with the membrane.

Trypsin is especially useful in cases where acid media
is not admissible, and is to be chosen also in all situations
where the smallest possible bulk of solvent agent is
desirable.

Trypsin may be applied by insufflation, pure or mixed
with sodium bicarbonate—four grains Trypsin to one of
soda ; or may be taken up on a wetted brush or probang,
or mixed with water and sprayed ; Trypsin, gr. 15, soda
bicarb., gr. 5, water, f ℥ i, to be prepared fresh every
few hours, or chloroform or pure creosote, 4 drops, may
be added as a preservative. For further details, see surgical use of the digestive ferments, p. 83, *et seq.*

DIASTASIC ESSENCE OF PANCREAS.

(FAIRCHILD.)

THE MOST ACTIVE, RELIABLE AND AGREEABLE AGENT FOR THE DIGESTION OF FARINACEOUS FOODS.

This preparation has been made with the especial purpose of obtaining the *diastase* or starch-digesting principle in an active and agreeable form.

The need had been often expressed to us by physicians, of a purely *diastasic* preparation by means of which they might assist the digestion of starch without at the same time introducing other digestive agents, or in any other way interfering with the process of digestion.

In meeting these requirements, this Essence has, we believe, been found peculiarly serviceable. It acts upon starch with great energy and promptness.

Inasmuch as the diastase of the pancreatic juice acts upon starch in a manner precisely similar to that of the saliva, this Diastasic Essence may be confidently expected to compensate for insufficient salivary digestion.

For this purpose it should be given at meal time— either immediately before or with the food.

When the intestinal digestion of starch is at fault, it should be given an hour or so after food.

This Essence of Pancreas is gratefully aromatic and acceptable to the most delicate stomach, and will be found, therefore, more efficient and agreeable as a diastasic agent than the thick, sweet extracts of malt.

It will sometimes be advantageous to mix the diastasic essence directly with the foods, such as oatmeal, rice, etc., especially for children who, owing to defective dentition or ill health, evince difficulty in the digestion and assimilation of starchy foods at an age when it is desirable that milk should no longer be the sole article of diet.

The Essence should never be added to food when too hot to be borne agreeably by the mouth.

Usual dose, one or two teaspoonfuls.

PEPTONISING TUBES

(FAIRCHILD.)

FOR THE PREPARATION OF PEPTONISED MILK AND OTHER PREDIGESTED FOOD FOR THE SICK.

(EXACT SIZE.)

Each tube contains the exact quantity of Extractum Pancreatis (grains 5) and of Soda Bicarb. (grains 15), to peptonise one pint of milk.

These tubes of "peptonising powder" are the most convenient means for prescribing and using the Extractum Pancreatis for the purpose of peptonising *milk.*

By this means the peptonising powder is supplied in an accurate and portable form, secured from deterioration, and dispensed at a moderate fixed price.

Each package contains complete directions for preparing peptonised milk, beef, gruel and a great variety of foods for the sick by means of the Fairchild Practical Recipes.

The tubes can be sent by mail. Retail price, 50 cents per box of one dozen tubes.

FAIRCHILD'S "DIRECTION SLIPS"

FOR THE USE OF THE PHYSICIAN IN PRESCRIBING PEPTONISED MILK, BEEF, GRUELS, ETC.

For the convenience of the physician we devised these "direction slips" in small pads of proper size for the vest pocket. The pad contains a number of slips of directions for each sort of food—peptonised milk by the cold process, and for jellies, for punches, etc.; peptonised gruel, peptonised beef, junket, whey, etc.

By this means the physician is enabled to leave with the patient or nurse plain printed directions for the special food and method he may desire to order. These direction slips have proven very acceptable to the profession. We shall be pleased to send them by mail upon request.

PEPTOGENIC MILK POWDER

YIELDS A FOOD FOR INFANTS WHICH IN PHYSIOLOGICAL,
CHEMICAL AND PHYSICAL PROPERTIES IS ALMOST
IDENTICAL WITH HUMAN MILK, AND AFFORDS
A COMPLETE SUBSTITUTE THEREFOR
DURING THE ENTIRE NURSING
PERIOD.

By means of the Peptogenic Milk Powder and process, cows' milk is so modified and pre-digested as to conform remarkably in every particular to normal human milk, thus affording a "humanised milk," exactly suited to the functions of infant digestion, calling forth the natural digestive powers of the stomach and supplying every element of nutrition competent for the nourishment and development of the healthy nursing infant.

It is also a peculiar feature of this method that the milk may be given just that degree of digestibility suitable to especial requirements,—in cases of naturally feeble digestion and during the disorders of infancy.

The Peptogenic Milk Powder is put up in $1.00 and in 50 cent packages, and sold by the principal drug houses in the United States and Canada. Sample can of the Peptogenic Powder and pamphlet will be sent gratis upon request. Correspondence solicited.

TESTS FOR PANCREATIC PREPARATIONS.

The pancreatic juice and the gland itself are well known to be extremely subject to decomposition, hence the greatest care and skill are required in manufacturing available medicinal products from this source. The value of a pancreatic preparation must depend not only upon its digestive activity, but upon the character, the quality of the digested product it yields. A pancreatic extract may convert the caseine of milk into peptone, yet the peptonised milk be quite unfit for food, owing to the development of rancid fatty acids, giving the milk a peculiar repulsive odor characteristic of regurgitated milk from a sour stomach. A pancreatic preparation which produces such a result with milk, is plainly unfit for any medicinal use.

We have in the past not infrequently had occasion to examine such commercial " pancreatines." A good pancreatic extract should rapidly digest milk, beef, fibrin and all forms of starchy food,—should convert the caseine of warm milk into peptone without the development of any rancid flavor whatever. The action upon caseine may be taken as a sufficient test of the proteolytic power upon any proteid. The activity and quality of a pancreatic preparation may be readily tested in the following manner :

Put into a flask 15 grains of sodium bicarbonate and 4 fluid ounces of cold water, add 5 grains of Extractum Pancreatis, mix well and add one pint of milk warmed to 130° F. Shake well and place the bottle convenient for observation. At first there should be no foreign odor or taste imparted to the milk. In a few moments the milk will become of slightly grayish yellow color, which in ten minutes will be more marked and the milk thinner and of a distinct bitter taste, due to the conversion of the caseine. This taste, even when peptonisation is complete, is a pure bitter without any suggestion of fermentation or rancidity. By having another flask of the milk mixed with the soda and water without ferment, the progress of the digestion may be, by comparison, more readily observed. These physical changes of milk, during peptonisation, are so characteristic, that anyone familiar with the process, may very readily regulate the process accordingly. By withdrawing a small portion of the milk from time to time and adding a few drops of acetic acid, the conversion of the caseine may be tested,

by the character of the curd formed—from the familiar
tough caseine, to the light, flocculent precipitate, and the
final slight, scarcely perceptible, granular coaguli.

To test the diastasic property of a pancreatic preparation, prepare thick, gelatinous starch, by mixing a drachm
of arrowroot or starch with five fluid ounces of cold water,
and boiling well. To a fluid ounce of this mucilage (at
110° F.), add a grain or so of the pancreatic extract or a
few drops of a fluid product and stir well. The starch
should become almost instantly thin and fluid, like water,
showing the formation of soluble starch, which is gradually converted into dextrine and glucose. A product
which does not quickly liquefy thick, warm starch jelly
is worthless as a diastasic agent.

FAIRCHILD'S DIGESTIVE TABLETS.

A PORTABLE AND EXACT FORM OF DOSAGE OF THE
DIGESTIVE FERMENTS.

These tablets are unique in form, agreeable to the
taste and easily carried about in the pocket. They are
offered as a means of exhibiting the digestive ferments in
divided doses and at the particular interval after the ingestion of the food, which gives the most favorable condition for their action. The advantages of this method
of administration are apparent, especially in duodenal
dyspepsia. The tablets should preferably be swallowed
whole.

The various combinations are supplied in small vials
and it is recommended that they be prescribed in original
bottles. The directions of the physician will be affixed
by the druggist in place of our label, if so desired; but
it will be economical to the patient to order the original
vial. They are also supplied in large vials in any quantity desired.

PEPSIN TABLETS.

(FAIRCHILD.)

Each tablet contains one grain of our pure *Pepsin in Scales*, combined with acids and appropriate aromatics. Dose—one or two tablets immediately after eating and repeated when required.

These tablets afford a means of re-enforcing the gastric digestion at frequent intervals after the ingestion of food. The advantages of this method of administering the peptic ferment have been well advanced in an editorial in the *New Remedies*, from which we quote; and the need of an available preparation for the purpose having been urged upon our attention, we originated these Pepsin Tablets which have proven very useful and greatly appreciated.

"Still another fact exists, although it has apparently been lost sight of in practice, and is rarely or never mentioned by writers on disorders of digestion, viz.: that much better results will follow the administration of the pepsin in divided doses during the process of digestion, and at intervals of a few minutes, than when it is given in one dose. The reason for this is the fact that as peptones are formed in the stomach, they are absorbed or passed through the pylorus into the intestine, and carry with them a certain proportion of the ferment which produces this change, and that in a case where the gastric juice is of notably poor quality, and artificial pepsin is employed, the digestive action, which at first may be quite efficient, grows weaker and weaker, and fresh supplies of pepsin are required from time to time to maintain the process. * * * "—*New Remedies*.

PEPSIN AND BISMUTH TABLETS.

(FAIRCHILD.)

Each tablet contains one grain of pure Pepsin (Fairchild) and two grains of Bismuth Subnitrate.

Pepsin and Bismuth constitute one of the most efficient and generally used combinations in the treatment of dyspepsia. In these tablets these remedies are presented in an exact, agreeable and efficient form.

The well-known chemical incompatibilities between Pepsin and Bismuth, in solution, and the criticisms justly urged against such a combination, have led some to the impression that this objection is true of Pepsin and Bismuth mixtures generally.

There is no question of incompatibility between Pepsin and Bismuth, except as relates to the Ammonia Citrate in solutions, a salt of Bismuth, moreover, which is greatly inferior to the Subnitrate. The Bismuth Subnitrate is well known to be very beneficial in certain forms of dyspepsia, and its properties are in no way inimical to the action of the gastric juice or to that of artificial peptic agents administered in conjunction with it.

Usual Dose.—*One or two tablets immediately before or after each meal, or at any time when suffering from indigestion.*

PEPSIN, BISMUTH AND PANCREATIC TABLETS.

(FAIRCHILD.)

Each tablet contains—Pepsin (Fairchild), 1½ grains, Ext. Pancreatis (Fairchild), 1½ grains, Bismuth Subnitrate, 2 grains.

Usual Dose.—*One or two of these tablets should be taken either shortly before or after meals, as may prove best suited to the particular case.*

"PEPSIN AND PANCREATINE" TABLETS.

(FAIRCHILD.)

Each tablet contains—Pepsin (Fairchild), 2 grains, Extractum Pancreatis (Fairchild), 3 grains.

This formula has been prescribed for some years by physicians of this city under the name of "Pepsin and Pancreatine," and we have supplied them uncoated for dispensing. The increasing demand made it necessary for us to prepare them in a manner uniform with our other digestive tablets, in order to permanently protect them from change.

The coating is perfectly soluble, and does not interfere with their digestive action.

One tablet, three times a day, is generally prescribed as a dose.

PEPSIN AND DIASTASE

(FAIRCHILD.)

IN TABLETS, EACH CONTAINING TWO GRAINS.

This combination, which is original with us, is the only preparation in which the pure diastasic and peptic ferments have, we believe, been united in an active form. The value and appropriateness of this combination is apparent.

It is certainly in clearest accordance with physiological principles. It is a well-ascertained fact that *diastase*, whether obtained from saliva, the pancreatic juice, or from germinated grain, acts upon starch in an identical manner and under identical conditions. "Pepsin and diastase" may, therefore, be given with every anticipation of beneficial results in cases of dyspepsia, where both the salivary and gastric digestion are at fault.

This combination is prepared with our pure pepsin, without admixture of malt sugar, starch or other substance, and in such a manner that an ordinary dose contains an efficient proportion of the diastasic ferment.

This product is not to be classed among those saccharated "digestive compounds" which purport to contain "all the agents of digestion "—diastase included. It is sufficient here to say that not one of them contains an appreciable quantity of diastase from any source. If a preparation contains active *diastase*, it must liquefy gelatinous starch at the temperature of the body.

One or more tablets for a dose at meal time, or when suffering from indigestion.

PEPSIN, BISMUTH AND NUX VOMICA TABLETS.

(FAIRCHILD.)

Each tablet contains Fairchild's Pepsin, 3 grains, Bismuth Subnitrate, 2 grains, Extract Nux Vomica, ⅛ grain.

COMPOUND OX GALL TABLETS.

(FAIRCHILD.)

Each tablet contains—Inspissated Ox Gall (Fairchild), 2 grains, Extractum Pancreatis (Fairchild), 2 grains, Extract Nux Vomica, ⅛ grain.

These two combinations having been much prescribed, we have manufactured them in our tablet form by request.

PANCREATIC TABLETS.

(FAIRCHILD.)

Each tablet contains 3 grains Fairchild's Extractum Pancreatis.

COMPOUND PANCREATIC TABLETS.

(FAIRCHILD.)

This tablet, originally designed for the treatment of intestinal indigestion, has proven of great service and has

been for some years extensively prescribed. The pure
Extractum Pancreatis is here combined with bismuth subnitrate, highly esteemed in allaying irritability of the alimentary tract, and with ipecac, which, in small doses, is
the most admirable stimulant of the intestinal digestion.

Each tablet contains—Extractum Pancreatis (Fairchild), 2 grains, Bismuth Subnitrate, 3 grains, Powdered
Ipecac, $\frac{1}{10}$ grain.

One or two tablets for a dose, an hour or two after eating.

PEPTONATE OF IRON
(FAIRCHILD),
IN TABLETS, EACH CONTAINING THREE GRAINS.

*Dose for an adult, usually one tablet thrice a day after
meals.*

FERROGLOBIN TABLETS.

Ferroglobin contains the element iron, united with
the proteid matter, as a constituent of the molecule itself,
thus presenting this important principle in a form peculiar
to the blood and impossible to produce artificially. Ferroglobin, therefore, may be considered to offer many advantages over any chemical compound of iron or any of the
mixtures of iron and albumen. Ferroglobin, in distinction
from all such artificial compounds, presents the organic,
physiological ferruginous element of the blood. It is
recommended in all anæmic conditions where it is desired
to administer iron in a perfectly soluble and assimilable
form. It is prepared with the utmost care and offered in
tablet form as the most permanent and acceptable preparation for medical use.

Each tablet contains 2 grains of pure Ferroglobin.

THE PEPTONISING PROCESS.

To peptonise food is to artificially digest food, to submit
it to the action of the digestive ferments, by which means

changes are effected precisely similar to those which in the living body are the essential preliminary to its absorption. For the two great types of food stuff, flesh and starch foods are incapable of being absorbed until they have become soluble by the action of the digestive juices, and thus capable of passing through the walls of the alimentary canal. This characteristic action of the digestive ferments, the conversion of insoluble and unabsorbable substances into soluble and assimilable, is seen in the artificial digestion of food.

The fibre of beef is seen to gradually soften and dissolve; thick, well-boiled, gelatinous starch (gruel) is seen to quickly dissolve, become thinner and watery. Farinaceous foods as ordinarily prepared, such as oatmeal, wheaten grits, rice, dipped toast, more slowly soften and dissolve. Albuminous substances, such as the caseine of milk, etc., acquire when completely digested, a bitter taste from the peptone; the farinaceous foods become sweeter from the maltose or starch sugar.

Cooked food is in general more susceptible to digestion than raw food, both in the body and in the flask. To peptonise food is then but to go a step beyond what has always been sought, in the special care and devices given to the cooking of food for the sick.

Each ferment has its special correlated food substance and this it will digest in a flask, just as in the alimentary canal. In discussing the ferments in detail, we have already had occasion to point out that pepsin is not available for household use in artificially digesting food of any kind. Peptonised food is, therefore, not food prepared with pepsin, or indeed necessarily containing a ferment of any kind; it is digested food; the agent of digestion may or may not be retained in an active form after its work has been utilised.

The pancreatic ferments are capable of digesting every known form of food ; and as made available in the Fairchild *Extractum Pancreatis, Peptonising Tubes* and other special forms, may be applied with marvelous facility for peptonising food for the sick by the Fairchild process, with the ordinary conveniences of the sick room.

The peptonising action is most energetic at about the heat of the body, slow at the temperature of a room (60 to 70 degrees F.) ; at a lower temperature, even at freezing, the peptonising agent is not destroyed, but is simply inactive. At the boiling heat it is at once killed.

Therefore we may peptonise milk by the cold process, in which the major work of the peptonising agent is done after the milk is taken into the stomach ; or by the warm process in which the milk is partially digested and then cooled to check digestion ; or after peptonising to a certain point the ferment is to be destroyed by boiling.

This boiled or scalded peptonised food contains now no active ferment, no artificial help to digestion ; we have removed the food from further influence of the peptonising agent, just as we remove food from the fire after cooking.

It will be seen, therefore, in the Fairchild's practical recipes that we have various simple methods, according to the degree of peptonising required, to suit the conditions of a case.

The effects of the peptonising process are as plain to sight and to taste, as are the effects of cooking and afford as simple evidence by which it may be regulated. It is in truth easier to tell when a pint of milk is peptonised to suit a given case, than it is to tell when an egg is boiled "soft," or "well done" or when a steak is properly broiled.

The peptonising powder always acts uniformly under given conditions ; those conditions are exceedingly simple and attainable.

It is of the greatest importance at the beginning, to follow the directions to the letter. With familiarity with the process, with its effects, with a clear idea as to the conditions essential and the object to be accomplished, then one may take one's "own way" to reach the desired result, to please, or agree with, any patient.

For instance, if peptonised milk should be required in an emergency, the powder may be mixed in a saucepan with warm water and warm milk and kept warm over a fire, say for five minutes, stirring briskly, and sipping frequently so as to take care that the milk is not overheated and the ferment thus destroyed. Thus in a few minutes peptonised milk may be so prepared as to be of the utmost service in affording urgently required absorbable nourishment. It may be given hot, or if required cold, ice it. Sometimes it will be found that the milk will agree (when made by the warm process), if it is put on ice the moment the milk becomes warm in the bottle, because the milk thus becomes sufficiently peptonised before it becomes chilled.

There is an exaggerated notion of the "trouble" of the peptonising process, probably because of the novelty of this application of physiological principles. But it is in reality an exceedingly simple process. It would be difficult to instance any of the commonest cooking operations so simple as mixing a powder, water and milk together and keeping it in a warm place (water bath or other) for a few minutes. If soluble, easily digestible, absorbable food is, as by all conceded, the chief desideratum, how shall we so simply, surely and safely obtain it as by the peptonising process? It is certainly far easier to peptonise food than to prepare most of the jellies, beef teas and delicacies in old time vogue for the sick. The peptonised foods have saved more lives in the ten years in which the Fairchild process has been in use, than all the other kinds of special foods for the sick that are made. It is scientific, practical, successful. If it chances that at the first attempt or occasionally, the milk becomes "too bitter," surely this is no more reason for condemning the process or rejecting it, than it would be to reject cooking because of the even greater difficulty of boiling an egg "twice alike" or of roasting meat "to a turn."

THE USE OF SODA IN THE PEPTONISING PROCESS.

As we have already explained, the use of an alkali is not essential to the action of the pancreatic ferments.

In the digestion of milk by the peptonising ferment the caseine undergoes gradual conversion, and at a certain point acquires the peculiar property of coagulating at the boiling temperature.

The caseine at this stage of its conversion is in the condition most generally suitable for digestion in the stomach ; it is no longer caseine and does not act like caseine and yet not completely peptone ; for peptone does not coagulate when boiled. It is in fact a peculiar partially transformed albuminoid which has been called meta-caseine, and it has been found that this may be prevented from coagulating from boiled milk by simply rendering it alkaline. Consequently by the use of a small quantity of soda bi-carbonate we are enabled to boil the milk, and thus check digestion at any requisite stage without coagulating the altered caseine. It is seldom necessary to

boil peptonised milk for adults except under circumstances when ice is not available to check the peptonising action. This addition of soda is also wholesome, it
neutralises the almost invariable acidity of cow's milk and
keeps it sweet.

In the Fairchild process for peptonising milk we direct
that the peptonising powder shall first be mixed with
water and then added to the milk ; the object being to so
dilute the milk that it will not be curdled by the digestive
agent. The action of the pancreas curdling ferment,
which we have described on page 34, is a hindrance to the
artificial digestive process ; for milk will peptonise more
readily, be more convenient for use, if kept fluid by the
simple expedient of diluting it with a small proportion of
water.

THE REASON FOR DILUTING MILK IN THE PEPTONISING PROCESS.

The Extractum Pancreatis contains the ferment of
the pancreas which curdles milk. This ferment does not
act well with diluted milk. Consequently, by the simple
expedient of adding a small proportion of water, we are
enabled to use the requisite quantity of Extractum Pancreatis to peptonise milk, without any interference from
the curdling ferment.

The addition of water in this proportion is not in the
least objectionable. For the great majority of cases in
which peptonised milk is resorted to as a diet, the addi-
tional water is a distinct advantage, for here it is that a
fluid food is of the utmost importance. It means that the
patient takes with every pint of milk four ounces of water,
and water is, in fevers, etc., the very thing required. It is not
so much concentrated food, as assimilable comprehensive

nourishment that is essential. In the special process for peptonising milk for infants, we direct the definite dilution necessary to yield a food containing the proportion of water found in human milk.

USES OF PEPTONISED FOODS.

It is no longer necessary to adduce "clinical experience" in support of the value of peptonised foods. Since we first had the honor to call the attention of the medical profession to the "Use of the Extractum Pancreatis in the Preparation of Peptonised Foods for the Sick," these foods have quite fulfilled their great promise of usefulness.

In a word, when the physician finds nutrition to be a factor in the treatment of a case, this is where peptonised foods are his chief resource. In peptonised milk, beef, gruels, etc., by the Fairchild process, the physician finds the food for the sick at once the most useful, economical, and congenial to direct for his patient. For the foods peptonised are the foods with whose composition and special nutritive properties and value, he and all mankind are familiar.

If the digestive functions are impaired, or even completely in abeyance, what other method of supplementing them so certain and so innocent? It is the most rational conceivable resource to thus accomplish digestion by proxy—to the degree only necessary to render the food *assimilable.* With returning health, the patient neither desires nor requires peptonised foods. The use of peptonised food reduces to a minimum the inroads which acute and wasting diseases make upon the system. The physician rationally anticipates a better convalescence, a quicker renewal of normal digestive power, for that patient whose nutrition has suffered the least degree of impairment. Let anyone compare the average results of the treatment, for instance, of "Typhoid" with the use of peptonised milk, with the results under the use of any other food.

There is the record of ten years, of innumerable cases in the use of veritable peptonised foods by the Fairchild process, without the citation of a single case of unfavorable sequelæ attributable to the use of these foods.

In experiments also with animals fed upon peptonised foods, there is no evidence of inability to return readily to the digestion of ordinary food. There was on one side a mere unsupported conjecture as to what might be the effect of protracted feeding of peptonised foods ; on the other, we know its beneficial effect after ten years of experience as a therapeutic resource ; we know that we do with unqualified benefit to the sick, subject food to preliminary digestion, and thus set disease at defiance in so far as it affects the most vital functions of digestion and nutrition.

That peptonised milk is competent for the complete nourishment of adults in *active life* suffering from gastric ulcer, or subject to chronic diarrhœa, is abundantly proven, and many instances have occurred in the past ten years where patients have found this their only resource for nutrition. We especially call attention to these typical cases, the personal experience of practicing physicians who have been enabled to pursue their profession and maintain vigorous life by subsisting solely on peptonised milk for years, under circumstances where otherwise "life had been a burden " from suffering.

" The history of this case of Acute Dysentery which " had progressed from acute suffering to exhaustion, emacia-

"tion and hopelessness ; which was not permanently bene-
"fited, but only controlled by the numerous drugs used
"against it, and which was at last cured by a simple diet
"of pre-digested milk rigidly adhered to by the help of
"obstinate will power, has appeared to me unique and
"therefore of use to the profession at large. It demon-
"strates that by recourse to the artificial process of diges-
"tion, we may present proper nutriment to our patients
"under conditions so unfavorable even as to render futile
"all other therapeutic measures, climatic and medicinal.
"It proves further, as will be shown, that under a pro-
"longed exclusive diet of so fluid an aliment as milk diluted
"with water and with its caseine converted into soluble
"peptones, health and activity may be maintained."

"Upon this diet, of milk peptonised by Fairchild's
"method, the patient has now been living exclusively for
"more than two years. His general condition is excellent.
"The functions of the bowels are performed with ease
"and regularity, his muscular system has regained its
"former degree of average development, and he bears,
"with the same ease, as do his fellows, the fatigues of
"either business or pleasure."

"As a physician of seventeen years of active practice,
"I have fully convinced myself of the great value of your
"'Pure Digestive Ferments'—particularly your 'Ex-
"tractum Pancreatis.' But my most valuable experience
"has resulted from my own personal experience. For
"three years I have suffered with gastric ulcer and
"Chronic Gastritis with frequent acute attacks. Life was
"intolerable and seemed about to terminate when I began
"washing out my stomach with medicated water and
"resorted to a diet of 'peptonised milk.' For 37 months
"I have lived absolutely upon peptonised milk, porridge
"and gruel."

In acute and wasting diseases, Typhoid, Pneumonia, Gastric Ulcer, Diabetes, Tuberculosis, Chronic Diarrhœa, Pyloric and Intestinal Obstruction, Gastric Catarrh, etc., as a food both preparatory and subsequent to important surgical operations, peptonised milk, gruel, etc., are the classical resource. In times of great fatigue and nervous prostration, when the strength is exhausted by severe strain of work and anxiety, when, as it is expressed, one is " too tired to eat," then peptonised milk has the most remarkable restorative power. See " Hot Peptonised Milk." (*Practical Recipes.*)

In Typhoid Fever, peptonised milk promises, and proves the "ideal food ; " it precludes all accumulation of unassimilable matter in the digestive tract and meets every requirement. It affords also the best vehicle and the most agreeable for the exhibition of the spirits, whisky, brandy, etc.

We do *not* recommend peptonised milk for feeding *nursing* infants, nor the use of the *Peptonising Tubes*, for preparing peptonised milk for infants. In "peptonised milk" (with the tubes), there is no attempt to adjust the milk quantitatively to a correspondence with human milk, nor to attain the definite proper conversion of the caseine. There is no reason, therefore, for using for an infant this method of peptonising milk designed for adults, when in the *Peptogenic Milk Powder* the process is, in every detail, adjusted to the analysis of normal human milk.

Humanised Milk as prepared with the Peptogenic Milk Powder is frequently preferred by the physician as a food for *adults*—in phthisis, Bright's disease, etc., because it is so fluid and agreeable, and yet richer in nutritive matter than pure cow's milk or peptonised milk.

PEPTONISED MILK.

Peptonised milk, some ten years ago practically un-known, is to-day by far the most important and the most used by the medical profession of all foods for the sick. The reason for this is shown in the great value of milk as a comprehensive nutrient, in its availability and cheapness. In truth, a pint of peptonised milk contains more actual peptone, more total nutritive substances, than the same bulk of many so-called concentrated beef elixirs, wines, etc., which cost a dollar per pint. Milk contains every element of nutrition in a form naturally fitted for absorption, with the exception of its caseine. Therefore, it is apparent that by changing the caseine into soluble peptone, we obtain an ideal food for the sick.

Caseine is, of all albuminoids, the most difficult and impracticable of artificial digestion by pepsin and acid, either as existing naturally in milk or as separated therefrom by acid or rennet, and treated just as we should treat egg albumen, fibrin, etc. Caseine is, moreover, unquestionably more difficult of digestion, even in the stomach, than other albuminoids, such as of fish, beef, egg, etc.

It is not a little remarkable that milk, "the type of a complete aliment," should prove so marvelously susceptible to artificial digestion by means of the proteolytic ferment of the pancreas, for thus the caseine of milk can be at will brought to any desired degree of conversion without rendering the milk repulsive in taste or appearance. In fact, peptonised milk, when prepared according to the directions with the peptonising tubes, is quite as agreeable as raw milk, and better relished by most persons. By the "cold process" no artificial taste whatever is imparted to the milk. Peptonised milk is milk with its caseine converted into peptone by the process of artificial digestion. The object of the directions given for peptonising milk is to submit its caseine to the action of the digestive ferment under the definite simple conditions by which it may be digested to any suitable degree of conversion.

When well peptonised, the milk will be found to have become thinner and of a greyish yellow color, and to have a slight, peculiar and by no means disagreeable taste characteristic of peptonisation. Wholesome peptonised milk should not have the slightest rancid flavor or odor.

It is very seldom necessary to peptonise the milk to the point at which the bitter taste is developed. It must be borne in mind that the peptonising process goes on as long as the milk is warm; therefore it is necessary to transfer the bottle promptly from the warm bath to the ice chest, in order to check digestion. For various methods of preparing peptonised milk, for making it an agreeable beverage, see Fairchild's " Practical Recipes."

NUTRITIVE ENEMETA.

Peptonised Foods are peculiarly adapted for rectal alimentation. In the rectum is presented every condition essential to the conversion and complete absorption of the peptonised food, etc., without irritation or complication. In times past it has been recommended to prepare beef, etc., for enemas by mixing it into a pulp with the fresh pancreas gland. To-day, in the Extractum Pancreatis or the Peptonising Tubes, we have the means of quickly and conveniently preparing milk, beef, eggs, etc., as absorbable enemas capable of sustaining life for an indefinite time.

MILK ENEMETA.

Milk may be introduced as soon as it is mixed in the ordinary proportion with the peptonising powder, and as it is usually required warm, a very considerable degree of pre-digestion will take place whilst bringing the milk to proper temperature ; or best, the powder should be mixed with ready warmed milk.

Peptonised milk may be very conveniently prepared by the cold process, and when required the proper quantity may be warmed and injected.

EGG ENEMETA.

Dissolve the white of an egg in thrice its bulk of warm water ; add 'the contents of a peptonising tube and stir well, and inject at once. An egg, white and yelk, may be thoroughly mixed with a pint of milk and peptonised in the usual manner, and thus afford a very nutritious enema.

BEEF ENEMETA.

Take a tablespoonful of minced lean beef, add to four tablespoonfuls of cold water, and gradually heat to boiling. Now rub all through a fine sieve or colander, and when luke-warm add the contents of a peptonising tube, and it is ready for injection. It may be made more fluid if desirable.

PANOPEPTON.

BREAD AND BEEF PEPTONE.

Having been the first to realise the value and scope of the digestive ferments as artificial agents of digestion, and the originators of the Fairchild process, which has become familiar in every household for the peptonisation of food for the sick, we have not failed to perceive the great need for a true, ready-made peptonised food. Peptonised foods by the Fairchild process have long been recognised as superior to all others available, the only objection being the necessity of preparing them fresh every day when required.

In Panopepton we present to the profession a new, complete and perfect peptone, one which we are confident will meet every requirement. Panopepton is the entire edible substance of prime, lean beef and best wheat flour, thoroughly cooked, properly digested, sterilised and concentrated in vacuo. The trimmed and cooked beef is subjected to digestion strictly to the point of complete solution of its albuminoids and the cooked wheat to the solution of both its gluten and starch. Panopepton is, therefore, the quintessence of peptones, containing all the nutrients of these two great types of food, beef and bread, fused into a delicious restorative.

The superiority of peptones from cooked foods over any form of raw, unsterilised beef is obvious. Sterilisation is an essential feature of the process for Panopepton, peptones not being coagulable at the boiling temperature, as are all other forms of albuminoids. Expressed juice of beef is instantly coagulated by heat, showing the fact that its albuminoids require conversion into complete solution before they are fit for absorption.

Panopepton is completely soluble and absorbable and responds to every test of true peptone and will satisfy the most exact and scientific scrutiny as to its qualities in every particular.

As significant of the technical skill and care with which the Panopepton is prepared, we call attention to the important fact that it is free from cane sugar or condiments, its agreeable flavor being purely characteristic and like that of roast beef juice and crust of bread.

Digestion is a process of solution, the slight mechanical operation concerned, being merely to expose increased surface to the solvent action of the digestive juices. By digestion only are we enabled to convert into solution the bulk, the actual substance, of food stuffs and thus fit them for appropriation by the system. We cannot by maceration or infusion with water, dissolve or extract the real nutritious substance of beef. The starch (the carbohydrate) of flour or bread, likewise, can be made soluble only by digestion.

Panopepton contains not only such extractives, salts and savory matters, as are found in beef juices, beef tea, etc., but further and peculiarly, a solution of the *whole substance* of beef and bread.

For many years the peptonisation of beef and wheat has been the subject of experiment and study by us, for we considered that in these combined albuminoids and carbohydrates only could we seek for a true and complete food.

If, for the nutrition of the body in health, every form of alimentary substance is essential, why should we in disease resort solely to albuminoids or digested albuminoids, except in the cases where especially indicated. The dietary experience of the human race is expressed in the saying, " bread is the staff of life."

The rank which peptonised milk holds as a food for the sick is due especially to the fact that milk is the "type of complete aliments."—(Dujardin-Beaumetz) ; "complete in itself."—(Pavy). Panopepton is the first food for the sick which may be relied upon to replace milk, for like milk, it affords all the elements requisite for the nutrition of the body.

The uses of such a peptonised food product as Panopepton are so obvious that it is only necessary to suggest the directions in which it will be found of inestimable value. Panopepton is the food par excellence, for invalids; in all acute diseases, fevers, etc.; in convalescence; for the large class of persons who from feebleness, or deranged digestion, or antipathy to ordinary foods, require a fluid, agreeable and quickly assimilable food. As a restorative from fatigue, for sleeplessness due to care and anxiety, or stress of mental work, Panopepton is a most potent reconstructive, to which immediate response is felt. Panopepton is preserved in a sound sherry, without added alcohol, and is at once a grateful stimulant and food.

A wineglass of Panopepton, with a small biscuit or cracker, will be found the best lunch or supper for the brain worker, when too tired for the tolerance or digestion of ordinary foods. For invalids travelling and under any circumstances where it is inconvenient to prepare food for the sick, Panopepton may be relied upon. In seasickness it is especially acceptable. Panopepton will be found to be of the most agreeable flavor when taken cold, consequently we recommend keeping it in a cool place, although it will keep perfectly for an indefinite time under ordinary conditions.

Panopepton should not be mixed with milk or any other food, and whatever diet is ordered in conjunction therewith, the Panopepton is to be taken pure or diluted only with ice-water, carbonic water or wines.

For Infants during summer complaint, Panopepton is a food rationally indicated and may be given in doses from a few drops to half-a-teaspoonful according to circumstances.

For adults the usual portion should be a small sherry glassful several times a day and at bedtime.

THE SURGICAL USE OF THE DIGESTIVE FERMENTS.

It is a fact long known, that the action of the proteolytic ferment of the gastric juice is not confined to purely alimentary substances, but is capable of dissolving albuminous matter in the various forms occurring in false fibrinous membrane, in sloughing and diseased tissues, etc. "Gastric Juice was many years ago employed by Dr. P. "S. Physick, the celebrated surgeon of Philadelphia, with "considerable success, as a local application to cancers and "sloughing ulcers, with the view of removing the dead bone "and flesh, correcting the offensive odor, and yielding a "healthy stimulus to the diseased surface. It has also been "used with success by Dr. Ellsworth, of Hartford, Conn., for "dissolving a portion of tough animal food, which had "become impacted in the œsophagus of a lad affected with "stricture of that passage. The gastric juice of a pig was "used.—(*Boston Med. & Surg. Journ.*, April 17, 1856.)

But the purely physiological functions of the digestive ferments and their application as agents of digestion of alimentary substances, have naturally more engaged the attention both of the medical profession and those who have sought to perfect the means and the method of utilising them in this most practical direction. Meanwhile the surgical application of the digestive ferments has too long failed of that attention which the least sanguine estimate of their value must show them eminently worthy of. In recent years we have given considerable attention to this subject, and from time to time supplied the ferments in the best form available for this purpose. Medical literature shows the record of the successful use of Fairchild's pancreatic extract and pepsin in the throat, in the auditory canal, in ulcers, sloughing wounds, in the bladder, etc. A novel and most important application of the digestive ferments in gonorrhea and urethral stricture has been made by a physician who has found Extractum Pancreatis the most successful agent. Applied dry it adheres to the mucous membrane, and finds sufficient moisture for its effective action. The Extract is preferably mixed with sodium bi-carbonate—say 1 grain to 5 of the Extract. In this situation, as in all others observed, the ferment seems to exert no action upon normal tissue. A record of many cases has already been made, and the subject is still under investigation. A case has also been reported to us of the successful treatment of stricture of the œsophagus by application of the Extractum Pancreatis with sodium bi-carbonate. In diphtheria, we have every reason to believe that the best results are to be obtained by the insufflation of the dry Extractum Pancreatis mixed with Sodium Bi-carbonate; thus applied, it adheres well to the mucous membrane, which affords sufficient secretion as a media for the solvent action. Dr. Robert T. Morris, a well known surgeon of this city, seeing the remarkable and signally successful use of Fairchild's pepsin, by his suggestion in the treatment of a crushed liver, was led to undertake scientific investigation and extended practical trial of the digestive ferments as solvents in surgical cases. Subsequently Dr. Morris undertook a series of experiments, to test the solvent power of pepsin upon carious bone previously decalcified by subjection to dilute Hydrochloric acid, with the view to remove dead bone without subjecting a weak patient to a dangerous or deforming operation. Succeeding in these experiments, Dr. Morris made practical use of this new surgical resource with complete success, the pepsin liquefying the carious bone and exerting no action upon normal bone.

The results of Dr. Morris' investigations are published in the *New York Medical Journal*, April 11th, 1891 : "The action of pancreatic extract and pepsin upon sloughs, coagula and muco-pus;" and March 19, 1892 : "The removal of necrotic and carious bone with hydrochloric acid and pepsin."*

The grounds on which the digestive ferments are applied in surgery are admirably stated by Dr. Morris as follows :

" It is not easy to see at a glance the whole field for digestive ferments in surgery, but we know that they are bland and harmless in any proportion, and that they will liquefy dead tissues close down to the living ones, and that there their action will end abruptly."

From Dr. Morris' paper the following typical cases summarised :

" A resource was brought into play a few weeks ago, when I had occasion to make suggestions relative to the treatment of a crushed liver. Portions of the organ, which were dark and sloughing, remained so firmly attached that their removal was dangerous, and the pultaceous lining membrane of the enormous abscess seemed to invite all manner of microbe guests. The idea of liquefying the dead tissues with a digestive ferment came into mind, and this being suggested. was carried into effect by the family physician, who injected into the abscess cavity a solution of scale pepsin, and, writing to me afterward, said : ' The pepsin did mighty good work. It broke up all dead tissues rendering them mostly liquid, and changed the color from brown to straw-color. The liquefied substances were easily washed out through the drainage tube. The wound was sterilised daily afterward with hydrogen peroxide, and the patient recovered without a bad symptom.' "

" Dr. C. N. Haskell liquefied two grammes of tough lining membrane from the tuberculous abscess of a case of hip joint-disease, with pepsin in fifty-five minutes."

" Dr. C. D. Jones, of Brooklyn, poured a solution of pancreatic extract (pancreatic extract, 2 dr.; water 8 ozs.) into the abscess cavity of a case of hip-joint disease one week after the operation of excision had been performed. He then wrote me as follows : ' The solution was allowed to remain in place half an hour, and the result was remarkable. Upon irrigation, I washed out numerous shreds of broken-down ligamentous tissue and many spicula of dead bone that had become imbedded in the

---

*Reprints of these papers will be sent on application to us.

soft tissues and that had previously escaped both irrigator and currette. The wound was then flushed out with hydrogen peroxide, and this treatment was followed by a marked improvement in the patient's general condition.'"

"In one case in which the bladder contained blood-clots and the catarrhal mucous membrane discharged ropy muco-pus, pepsin injected for the purpose of liquefying the clots not only fulfilled its mission in that direction, but unexpectedly cleared out the muco-pus and left the interior of the bladder quite clean. The process was repeated as soon as the muco-pus again became abundant, and the patient experienced a feeling of relief after the simple cleansing that pepsin afforded."

"After much experimentation I have finally adopted a method of work which seems to be complete. An opening is made through soft parts by the most direct route to the seat of dead bone, and if sinuses are present they are all led into the one large sinus if possible. The large direct sinus is kept open with antiseptic gauze and the wound allowed to remain quiet until granulations have formed."

"Granulation tissue contains no lymphatics, and absorption of septic materials through it is so slow that we have a very good protection against cellulitis. The next step consists in injecting into the sinus a two or three per cent. solution of hydrochloric acid in distilled water. If the patient is confined to bed, the injections can be made at intervals of two hours during the day ; but if it is best to keep the patient up and about, the acid solution is thrown into the sinus only at bed-time. In either case the patient is to assume a position favorable for the retention of the fluid. Decalcification takes place rapidly in exposed layers of dead bone, and then comes the necessity for another and very important step in the process. At intervals of about two days an acidulated pepsin solution is thrown into the sinus (I use distilled water, f $\mathfrak{Z}$ iv ; hydrochloric acid, m, xvj ; Fairchild's pepsin, $\mathfrak{Z}$ ss.), and this will digest out decalcified bone and caseous or fatty *débris* in about two hours, leaving clean dead bone exposed for a repetition of the procedure. The treatment is continued until the sinus closes from the bottom, showing that the dead bone is all out."

"Even in distinctly tuberculous cases the sinuses will close if apparatus for immobilising diseased parts and tonic constitutional treatment are employed, as they should be in conjunction with our efforts at removing the dead bone."

"If suppuration is free in any cavity in which we are at work, it is well to make a routine practice of washing out the cavity with peroxide of hydrogen before each injection."

Pepsin is the ferment which will probably give the best results in all cases where the acid essential to its operation is not objectionable, and where the swelling of the fibrinous matter, which instantly occurs on contact with the acid is not an objection. This behavior of acid would sometimes be considered unfavorable, as in the auditory canal, in the throat and in the urethra ; but in abscess cavities, etc., this is no objection—on the contrary, the acid seems itself a salutary agent, giving a healthy stimulus to the diseased surface, and is, moreover, antiseptic. Further, the action of this acid-pepsin digestion ceases at the production of peptone. These pepsin peptones do not readily undergo themselves putrefactive changes whilst the solution remains acid. The pancreatic ferments acting upon all these forms of proteid encountered in surgical cases, are very effective in water without the intervention of an alkali, but their action is accelerated in an alkaline solution so slight as one part sodium bicarbonate to 500 of water. Moist fibrin can as quickly be digested by Extractum Pancreatis plus alkali as with pepsin plus acid, by simply adjusting the quantity of the ferment to attain the desired result. There is no possible cause for hesitancy in using the ferment as freely as necessitated ; there is no other effect than its digestive action, which ceases when no morbid tissue remains to work upon. The soda itself is in many instances as clearly indicated and as useful, for instance in the urethra, in the bladder, in the throat, etc., as the acid is in the cases suitable for pepsin.

In the surgical use of the digestive ferments, it is absolutely essential to follow as closely as possible the conditions most favorable to the action of the particular ferment utilised. At the outset, probably no better guidance can be had than the procedure developed by practical experience in the use of the digestive ferments, as applied in this artificial digestion of albuminous matter in the test tube.

In using pepsin, the intervention of acid, from one half to one per cent. hydrochloric acid, U. S. P. to the volume of water is essential. Extractum Pancreatis may be used with simple water, or with water rendered slightly alkaline with soda bi-carbonate, say 5 grains to each fluid ounce.

The surgeon, then, in the normal range of media and action of the peptic and pancreatic ferments, is enabled to use an effective solvent, either acid, neutral or alkaline, as best adapted to the case in hand.

The most favorable temperature for the preparation and for the action of the digestive fluids can be readily ascertained by any attendant without the use of the thermometer, by using water heated to the point at which it can be borne by the whole hand (115° F.), or not too hot to be swallowed with comfort, about 130° F. This gives a temperature of 115 or 130° F., at which they act better than at the body heat, and the water should always be brought to the proper temperature *before adding the digestive ferment.* This avoids all risk of injuring the ferment.

If there is no cavity to hold the solvent in contact with the matter to be digested, the solvent should be applied by copious sprays, frequently repeated. In cavities, repeated applications are preferable, as otherwise the digestive fluid may become saturated with the products of digestion, and thus cease to act. It is also to be noted that the irrigation should follow as quickly as possible the liquefaction of the tissues.

The solvents should *invariably be freshly prepared for each application,* as the ferments mixed with the water are not only prone to decomposition and to become inert, but if mixed with cold water and then brought to the proper temperature each time required, they are very apt to be injured by overheating. With warm water, it is but a moment's work to prepare just the quantity required for each application.

In using pepsin, we strongly recommend *Glycerinum Pepticum*, both for convenience and efficiency, as containing the ferment in a highly concentrated, pure glycerin extract, instantly soluble in any desired proportion and especially convenient for spraying.

GLYCERINUM PEPTICUM
AS A SURGICAL SOLVENT.

In any convenient glass, mix one teaspoonful of Glycerinum Pepticum with one fluid ounce of warm water, say at 115° F., and 4 drops acid hydrochloric c. p. (16 drops dilute acid U. S. P.). Apply by injection, spray, etc., as most suitable.

FAIRCHILD'S PEPSIN
AS A SURGICAL SOLVENT.

Mix 5 grains of Fairchild's pepsin in powder perfectly smooth with a teaspoonful of water, then add an ounce of warm water, stirring well. Add 4 drops acid hydrochloric c.p. and apply as required.

EXTRACTUM PANCREATIS
AS
A SURGICAL SOLVENT.

Mix in any convenient glass, 5 grains Extractum Pancreatis to each fluid ounce of warm water, first carefully stirring the powder with a teaspoonful of the water, to a perfectly smooth mixture. At the option of the surgeon, soda bi-carbonate, about 5 grains to each fluid ounce, may advantageously be added.

It is by no means essential that these or any arbitrary proportions shall be observed. These quantities specified can be readily approximated without weighing. If it is found desirable to write a *prescription* for the solvents, the following formulas will be found satisfactory :

R̵   Glycerinum Pepticum..... .. ....2 fl. ozs.
          Acid Hydrochloric c.p...........64 minims.
          Aqua-Destillata.... ....6 fl. ozs.

M, S. Pour the quantity necessary for each application into an equal quantity of water, heated to about 115° F., or as hot as can be borne by the whole hand and apply as directed.

R̵   Extractum Pancreatis .... 3 drs.
          Soda Bi-carb. .... 1 dr.

M, Divide in wax papers No. 12.

Mix one powder with a gill of warm water and prepare fresh for each application as directed.

In preparing these solvents with water, it is well to use that which has been well boiled or distilled.

R̵   Extractum Pancreatis .... 1 dr.
          Soda Bi-carb.................15 grains.

M, Divide in wax papers No. 12. Apply *dry* as directed.

FAIRCHILD'S

PEPTOGENIC POWDER AND PROCESS;

ITS DEVELOPMENT AND RATIONALE.

It is now more than seven years since we introduced a method of preparing an imitation of woman's milk, based upon the agency of a digestive ferment in effecting the physiological conversion of the caseine of cows' milk into the soluble and diffusible form, in which the albuminoids exist in human milk.

This purely physiological action of the digestive ferment can be controlled or checked at will by regulating the temperature to which the ferment is subjected. Under favorable conditions, the ferment acts until its power is spent; by simply raising the temperature of the digesting mixture to about 160° F., it is instantly destroyed and thus becomes an inert substance, insignificant in amount and resembling in chemical and physiological properties so much albumen.

The digestive ferment has no other action, no properties at all comparable to those of a drug or chemical.

The pancreas ferment, trypsin, has a remarkable affinity

toward milk, digesting its caseine with great rapidity without altering its other elements and without rendering the milk repulsive. Milk so treated is known as "peptonised milk " and as made available by the *Fairchild products and process*, has long since become the chief reliance of the medical profession as supplying an ideal food for the sick.

The value of peptonised milk was immediately recognised as a resource for the feeding of an infant with naturally feeble or disordered digestion, and experience only confirmed the promise of its great usefulness.

After some years of practical experience with the peptonising process, and seeing the great facility with which caseine could be brought to any desired degree of digestion, the idea occurred to us that this process promised the solution of the problem of preparing an adequate substitute for woman's milk; as opening the way for the qualitative and quantitative adjustment of cows' milk to a correspondence with human milk to a degree never before attempted.

In order to prosecute the undertaking, we entrusted the necessary expert investigation to Dr. Albert R. Leeds, well known to have made especial study of the composition of human milk and of the infant foods.

Dr. Leeds found that under the influence of the digestive ferment, the caseine could be so altered as to impart a new and peculiar property to the milk ; that the milk became in its physical characteristics, density, color, taste, and in its behavior with acids and with gastric juice, remarkably like mothers' milk. The albuminoids of this converted milk under analysis showed a close resemblance to the albuminoids of woman's milk, thus adding a final proof to the theory that the differences in the physical properties and in the behavior, and digestibility of cows' and human milk, are directly dependent upon the character of their albuminoids.

Here then was an agent for the physiological conversion of the caseine, an expedient far more effective, natural and convenient than any hitherto devised, and which made possible the construction of an artificial human milk, by carrying out the further important modifications indicated by the results of comparative analysis.

As a final result, we were able to offer the Peptogenic Milk Powder and method which were found by Dr. Leeds, as stated in his report, "to yield a 'humanised milk' which in physical characteristics and chemical constitution approaches very closely to woman's milk." Thus we introduced a method of preparing a substitute for mothers' milk, which is the direct result of scientific study and investigation and which fairly represents the present status of knowledge and attainment ; the only food for infants which in its development and accomplishment, conforms to the universally accepted postulate that the best artificial food for an infant is that which in the highest degree resembles mothers' milk.

Now after seven years of practical experience, study and investigation since its introduction, during which time we have had abundant means of ascertaining the results of its actual use as an exclusive substitute for mothers' milk, we feel the strongest conviction that it affords an artificial food for infants which is entirely adequate for the nourishment and development of an infant during the nursing period. We have, during these years, given unremitting investigation in every direction which might enable us to attain the utmost perfection of detail in the approximation to the natural food of an infant.

We submit the Peptogenic Powder and method solely upon this ground,—as a scientific, practical and successful method of modifying cows' milk to the known composition

of human milk. Upon this ground, we ask the consideration of every physician interested in providing food for
infants deprived of breast milk.

THE POINT OF VIEW ON INFANT FOODS.

In infant feeding, as in many other subjects, scientific
standards are in advance of practical usage. So whilst
everywhere it is premised that mothers' milk is the best
food for an infant, we see that foods which are wholly made
up of substances foreign to milk, foods which were never
designed to resemble human milk, continue to be bought
and used without a question as to how they resemble the
food for which they are to be substituted.

We ask the physician therefore to submit the infant
foods of the shops to the practical point of inquiry.   How
do they resemble mothers' milk when prepared for the
nursing bottle?   No " infant food " as it is found in commerce, resembles mothers' milk, or can take the place of it.

The " infant foods " of commerce may be fairly divided
into two distinct classes :

First—Those which do not contain any milk and which
are to be made ready for the nursing bottle by admixture with cows' milk.

Second—Those which contain milk, the dried and condensed
milk foods, all of which are directed to be prepared for
the nursing bottle simply by the addition of water.

When mixed for use according to the "directions " of
the manufacturer the "infant foods" differ from human milk
obviously in physical properties, and by analysis will be
found invariably deficient in milk fat, milk sugar and milk
salts, which deficiency is not by any means compensated
for by the malt sugar or baked flour which imparts thick-
ness and sweetness to the food.

Another important question is :—Shall we use fresh milk or commercial milk products as the basis for the preparation of an infant food ?

From our standpoint, we can at present find no other basis than fresh milk, for we have not thus far found that milk can be so treated as to afford a stable commercial product, from which a close approximation to mothers' milk can be prepared.

We hold to the view that milk is materially altered by drying. That it will not on the addition of water, have restored to it even the physical characteristics of the original milk ; that the dried caseine of the milk will not again dissolve in the water ; that the milk fat cannot be dried successfully ; that it will in this state, soon become rancid.

It has for these reasons never been found possible to dry pure unskimmed milk as a marketable product, even for ordinary culinary and dietetic purposes.

Milk condensed without sugar or other preservative has been found apt to spoil quickly after the package is opened. Sweetened condensed milk contains a large amount of cane sugar and when such milk is diluted with the proper amount of water, it is much too sweet and thick. It is in common practice diluted so as to be greatly deficient in the real elements of milk.

COWS' MILK AS A FOOD FOR INFANTS.

In seeking a food for an infant deprived of breast milk, cows' milk has been instinctively resorted to as the substance nearest to it in apparent properties and design. But notwithstanding these similarities, cows' milk has proven inherently indigestible for the infant stomach and inadequate to replace woman's milk. Hence the problem of infant feeding. Hence the infant foods of commerce. Finding that cows' milk forms an indigestible curd, various expedients have been employed to overcome this. Among these, is the use of an alkali such as lime water, to form soluble or alkaline albuminates, to give the milk an alkaline reaction and also to retard the curdling action of the gastric juice.

The most familiar and common method has been to thicken milk with baked flour, or farinaceous foods—to keep the curds from forming a mass. But the effect of these substances is purely mechanical, they do not alter the character of the caseine, they " thicken," but do not enrich milk. It is simply adding a new difficulty without overcoming the original one.

Liebig, seeing that starch was not suited to an infant's digestion, that the nursing infant is not endowed with the power to digest starch, proposed to utilise the starch digesting ferment of malt—its diastase—for the artificial digestion of starch. Thus he gave us the method of treating wheat flour with a fresh infusion of malt and bicarbonate of potash, by which the starch is dissolved and converted into malt sugar. This, in brief, is the origin of the Liebig foods, by which we are now supplied with a ready-made malted or digested flour for addition to fresh milk.

But there being no starch nor digested starch (maltose, dextrins, etc.) in milk, human or animal, whilst there is in all milk found available a sugar peculiar to milk alone, there remains neither reason nor necessity for giving starch or digested starch to the nursing infant.

Therefore, from the standpoint of to-day, Liebig's method cannot, on theoretical grounds, be considered to afford an approximation to mothers' milk, nor has it in long practical experience proven a solution of the problem of infant feeding.

In the futile attempt to overcome the indigestibility of caseine, milk has often been so diluted as to render it incapable of properly nourishing an infant. The addition of water does not alter the character or behavior of caseine. Many in diluting the milk, have added nothing to attempt to compensate for the dilution of the sugar and the fat,— at the beginning deficient in quantity.

Such in brief, are the methods long in practice for the preparation of fresh cows' milk for infants. So that whilst the selection of animal milk for the bottle feeding of infants has been dictated by its resemblance to the natural food, we have gone on adding to it, substances foreign to all milk and unsuited to the digestive functions and nutrition of an infant.

It has taken us a long time to get to the present standpoint, that "an infant food approaches perfection in the degree in which it resembles human milk."

If many infants have, by virtue of superior resistance, been capable of appropriating sufficient nourishment from cows' milk in the various forms, how many have perished by artificial feeding! It is not here necessary to "make a case" in order to offer a remedy.

COMPARATIVE COMPOSITION OF COWS' AND HUMAN MILK.

Modern chemical and physiological investigation clearly reveals the reason why cows' milk is not suitable for the human infant. We see the significance of the difference found to exist in the composition of human and cows' milk, —that the milk of each is peculiarly adapted for the purpose for which it is designed. We now know that cows' and human milk differ in the total quantity of nutritious materials and in their relative proportions. Cows' milk contains less total solids, less fat, less milk sugar and twice as much albuminoids. In cows' milk, there is a larger proportion of the element of nutrition which creates and supports muscular energy and activity ; in human milk there is a larger proportion of sugar and fat.

Breast milk is uniformly and persistently alkaline. Cows' milk is more or less acid, and its acidity becomes more and more marked by keeping.

In cows' milk the greater part of the albuminoids is caseine, the substance which is curded by rennet and precipitated by acid—the cheesy portion.

In woman's milk the greater part of the albuminoids exists in a soluble or peptone-like form, which is incapable of coagulation or precipitation. The small fraction that is coagulable gives with acid or gastric juice minute, mobile, flocculent particles.

Thus it appears that in cows' milk there is not only a preponderance of albuminoids, but their *quality* is such as to demand a degree of digestive power to which the infant organism is unequal.

Milk is a vital secretion, and human milk the more highly elaborated, in its digestibility and its nutritive qualities, in conformity with the requirements of the highly organised being for whose nutrition it is destined.

It is the caseine, therefore, which has proven the obstacle to the practical employment of milk as a food for infants, for the sugar and the fat of milk exist in a form ready for absorption—and in a form peculiar to milk alone.

THE USE OF THE PEPTOGENIC MILK POWDER FOR THE PREPARATION OF "*HUMANISED MILK*" INVOLVES THREE DISTINCT STEPS :

First—To prepare with Peptogenic Powder, cows' milk, water and cream, a mixture which has the *quantitative* compo. sition of average human normal milk.

Second—To subject this mixture to the action of the digestive principle by which the albuminoids (caseine, etc.)
are converted into such form as to become identical
with those of human milk.

Third—To then destroy the digestive ferment by simply
raising the temperature of the milk to the boiling
point. This heat also destroys the bacteria and renders the milk practically sterile during the time
required for use—24 hours.

DIRECTIONS FOR "HUMANISED MILK."

No. 1.

FOR THE DAILY FOOD OF A HEALTHY NURSING INFANT.

Put into a clean granite ware or porcelain lined saucepan,
four small measures*, or one large measure of the Peptogenic
Powder, half pint of cold water, half pint of cold fresh
milk, and four tablespoonfuls of cream. Place the saucepan on a hot range or gas stove and heat with constant
stirring until the mixture boils. The heat should be so applied as to make the milk boil in ten minutes.

Keep in a clean, well-corked bottle in a cold place.
When needed, shake the bottle and pour out the desired
portion and heat to the proper warmth for feeding—lukewarm.

No. 2.

SPECIALLY PREPARED FOOD FOR INFANTS WITH FEEBLE
DIGESTION OR WHEN SUFFERING FROM DISORDERED
STOMACH AND BOWELS, AS IN CHOLERA
INFANTUM, ETC.

Put into a clean bottle, four small measures*, or one large
measure of the Peptogenic Powder, half pint of cold water,
half pint of cold, fresh milk and four tablespoonfuls of cream.

* Each large can of Peptogenic Milk Powder contains a large and a small
measure. Put the Powder into the measure with the blade of a knife, shaking it
down firmly so as to well and evenly fill the measure.
The small can contains the small measure only.

Shake well, place the bottle in a pail or tin kettle of water (at least a gallon) as hot as can be borne by the whole hand (115° F.), and keep the bottle there for 30 minutes. Then pour all into a sauce pan and quickly heat to boiling point with constant stirring.

Keeping and feeding in the same way as directed in No. 1.

COMPOSITION OF "HUMANISED MILK."

"Humanised milk" contains the amount of milk sugar, fat, albuminoids, ash and water found in mothers' milk. It possesses the peculiar alkaline reaction due to the proper proportions of those various mineral and saline constituents which are always normally present in woman's milk, and which are essential elements in the nutrition of the infant, being vitally necessary to the development of its osseous system. It resembles mothers' milk remarkably in its physical properties, and under every known method of test, it is found to behave in the manner characteristic of average normal breast milk.

We do not advise varying the proportions according to the age of the child. In the careful study of the facts brought out by the many analyses now extant of woman's milk, made during the entire period of lactation, there does not appear a sufficient variation in the quality of the milk, or in the ratio of its constituents, to afford a practical ground for making any variation in an artificial food.

It may logically be assumed therefore, that the average composition of human milk is the most practical and scientific basis for the fabrication of a food for the average infant, permitting the bottle-fed infant, like the nursing infant, to take food in such quantities and at such intervals as best conduces to its health.

DIGESTIBILITY OF "HUMANISED MILK."

"Humanised milk" presents to the infant's stomach a food which requires the same exercise of the natural digestive functions as required for mothers' milk—the caseine has undergone no greater amount of artificial digestion than is necessary to bring it to the soluble condition characteristic of the albuminoids of mothers' milk. It is not in the least giving a milk unnaturally easy of digestion. There enters the infant's stomach no artificial aid to digestion, no pepsin, no digestive ferment of any kind ; for after the ferment has accomplished a certain work in the conversion of the caseine, it is then destroyed and has no further influence upon the food—has nothing more to do with the digestion of the milk in the stomach than has the fire by which the milk was heated.

HOW TO ADAPT THE MILK FOR INFANTS WITH FEEBLE DIGESTION.

In order to adapt this "humanised milk" to the stomach of an infant with naturally feeble digestion, or with digestion disordered by teething, summer complaint, etc., the food is not to be specially diluted, it is simply necessary to regulate the degree of conversion of the caseine to insure its digestion and assimilation. This is accomplished by leaving the milk for a longer time at the temperature suitable for the action of the ferment before boiling the milk. Therefore in directions No. 2, we direct 30 minutes in the warm water bath before bringing the milk to the boiling point. In extreme cases, the caseine may be, by longer digestion, (40 to 50 minutes) so converted into a soluble form that the milk becomes capable of absorption without the least tax upon the stomach.

As the child recovers strength, the degree of this

conversion is gradually decreased until it is able to assimilate milk which has the digestibility of mothers' milk. By this means a sick infant is not deprived of nutrition in the attempt to find a food which it can tolerate.

There is also another expedient which has been found successful in the many cases of infants who seem to have practically no digestive power. This is to give the "humanised milk" containing the ferment in an active form and thus capable of effecting the subsequent changes of the food essential to its assimilation. Mix the Peptogenic Powder, water, milk and cream in the regular proportions, *cold*, then place the bottle directly *on ice*. When required, shake well, pour out only the necessary quantity and heat carefully over a flame until it is just warm enough for the nursing bottle. Do not boil it and do not let it get hotter than is agreeable to the mouth. This method should always be used when the food by Directions 1 or 2 is not properly assimilated by the infant. After the child has become strong enough, then gradually return to Directions No. 1.

CHOLERA INFANTUM.

The Peptogenic Milk Powder is too often brought first into use in a case when the infant is suffering from cholera infantum or from severe disturbances of digestion, aggravated by improper food and the system weakened by lack of nutrition. Even in such cases, relief is often immediately found in the administration of the specially prepared "humanised milk." (See directions No. 2). Give very slowly and in very small quantities at each feeding. But many times the case presents the entire alimentary tract in a condition highly favorable to the fermentation of milk and equally unfavorable to its absorption. Hence, milk may add fuel to fire.

In consequence of these facts it has become the practice to discontinue milk entirely for a time, in the endeavor to give rest to the digestive functions and to promote the effects of the purely medicinal measures. The difficulty here is to find a substitute for milk which will afford adequate nutrition.

WHEY AS THE TEMPORARY FOOD IN CHOLERA INFANTUM, ETC.

We strongly recommend " Whey " as affording by far the most satisfactory *temporary* food in Cholera Infantum.

This opinion is based not only upon its composition, but also upon some seven years of practical experience of its use. Whey contains, in a greater or less degree, every element of nutrition and in a perfectly assimilable form. It contains the soluble albuminoids, milk sugar and saline constituents of the milk. It is thus not giving a diluted milk. The milk has only been deprived of its caseine and the greater portion of its fat. Whey is therefore unquestionably far more suitable for the nourishment of a child than beef juices, beef foods, etc., or any other food which has ever been suggested as a temporary substitute for milk. As prepared with Fairchild's Essence of Pepsine, it is not only most palatable, but the contained Essence is also of great value as a remedy. For very young infants, it should be prepared with one teaspoonful of Essence to a pint of warm milk. For older infants it may be prepared with a teaspoonful to a half pint of milk and thus the proportion of the Essence of Pepsine may be made available for its remedial properties in conjunction with the whey. The Essence of Pepsine is accompanied by simple directions for the preparation of whey. Whey should be given from the nursing bottle, like the ordinary food.

The Essence of Pepsine affords the ideal digestive and carminative stimulant for disorders of infant digestion. It is generally given in 5 to 10 drop doses in a teaspoonful of pure water or with a teaspoonful of the whey or "humanised milk."

HOW LONG SHOULD THE BABY BE FED ON "HUMANISED MILK."

Only that food is a proper substitute for breast milk which is capable of the nutrition of an infant during the entire nursing period. "Humanised milk," being equivalent to average healthy breast milk, should be the exclusive food of an infant just as long as it would ordinarily take breast milk. The soundness of this theory and this practice has been proven by experience—by results.

It is found that there is a disposition on the part of parents to hurry the child along to what they fancy to be a "richer" food, to milk "thickened" with prepared foods, etc. To which we reply that "humanised milk" is not deficient in any element for the perfect development of an infant, that it is as rich as human milk, and richer than cows' milk in every constituent save the caseine; that mothers' milk ought to be the safest standard for a food up to the time of weaning.

HOW TO WEAN THE BOTTLE-FED BABY.

An infant should be weaned from the bottle gradually just as from the breast. At an age at which a nursing child would ordinarily be given a little oatmeal, hominy or rice, the bottle-fed infant should be given these farinaceous foods.

Begin with one feeding a day of well boiled oatmeal or rice, or some well baked potato mixed with "humanised milk." Feed with a spoon. Increase gradually to several times a day, or until the bottle is no longer required. Now begin to prepare the food with ordinary pure fresh cows' milk instead of with the "humanised milk" until you accustom the child to live entirely upon pure milk and farinaceous foods. But the less meat the better, until the child is two or three years old, say many of the physicians most expert in infant feeding.

"HUMANISED MILK"

AS A PARTIAL SUBSTITUTE FOR BREAST MILK.

In many cases it is found desirable or necessary to resort to bottle-feeding as a *partial* substitute for breast milk; here the "humanised milk" is the only food which can be properly given.

It is so identical with pure breast milk that no injury results to the child, it is taken as readily as the breast milk, and this alternate feeding produces no disturbance of the digestive functions. It is so much better than faulty breast milk, that it is often of the greatest value, both to the infant and mother to resort partially to a food which properly nourishes the child and relieves the mother of an undue tax upon her strength. No good result can come from compelling a child to take several times a day, thick, sweet malt sugar or starchy food, foods which load and distend the stomach, and the rest of the time the thin, digestible fluid—mother's milk.

Many a mother would gladly and profitably be relieved in a measure of the strain of nursing, if it could be accomplished without prejudice to the child, and this can be done by means of "humanised milk."

"HUMANISED MILK" HAS NO SPECIAL EFFECT UPON THE BOWELS.

The "humanised milk" has no especial tendency to produce either costiveness or looseness of the bowels.

Either one of these conditions may appear according to the constitution of the child or as dependent upon various reasons, just as may occur when taking breast milk. Sometimes—especially in hot weather—an infant requires *water* to keep its bowels in good order and for its well being in general. A little calcined magnesia or a little flake manna dissolved in the milk when ready for feeding, is a good remedy for constipation. Or use oatmeal water in place of plain water in preparing the "humanised milk." Take one table-spoonful of thoroughly cooked oatmeal (as ordinarily prepared for the table) and stir well into half a pint of hot water ; strain. Constipation is sometimes immediately relieved by heating the milk to 170° F. instead of to the boiling point ; this lower temperature is equally effective in killing the digestive ferment and in sterilising the milk.* Hold the fresh milk mixture in a saucepan over a flame, stirring constantly till it is heated to 130° F., then place over the flame again and stir constantly till the milk reaches 170° F.—the whole process not to take more than 10 minutes. Then pour the milk into a clean, well corked bottle.

One of the most simple and frequently effective expedients for loose bowels is to thicken the milk for a few feedings with thick arrowroot gruel, made by mixing the arrowroot with cold water and then boiling it for a long time till very smooth and well cooked. The so prepared gruel is to be added to the "humanised milk" when it is ready for feeding to the child. Its use should only be continued for a few feedings until the trouble is remedied. Colic, loose bowels with flatulence, are greatly relieved by the use of Fairchild's Essence of Pepsine given in from 5 to 10 drop doses in a teaspoonful of water just before

---

* For this purpose buy the cheap dairy thermometer, all glass and plainly marked to 170° F.

feeding. It may be so given several times during the day, but not continued beyond the necessity for its use. If there is persistent diarrhœa, it is a case for the physician ; it requires skilful medical treatment.

"CHANGING THE FOOD"

AS AN EXPEDIENT IN GETTING ONE THAT WILL AGREE.

It must be held that having a food equivalent to mothers' milk, we should use it like mothers' milk and as far as possible treat all variations of function and disturbances of health just as we would if the child were taking breast milk.

In the average healthy infant, fed from birth upon "humanised milk," there is as little probability of digestive disturbances as from breast feeding. Indeed, the "humanised milk," is often advantageously substituted for faulty breast milk and successfully *alternated* with healthy breast milk. But the difficulty is that foods are for the most part selected (?) hap-hazard—the food which "sells the most," or is the most advertised, or the food upon which a friend's child has been brought up, etc. The result we see is that in numberless instances coming to our knowledge, it is a history of one food after another, as many foods sometimes as the infant is months or even weeks old.

These cases present the utmost difficulty when medical advice is finally sought ; every possible variety of complication is encountered. In these cases, even good breast milk would not at once be successful.

An infant accustomed to the unnatural distention, irritation or stimulus to the alimentary tract from the presence of bulky, thick, insoluble " farinaceous" foods, or "milk foods" composed of dried milk and baked flour—in which both caseine and starch are in a practically unassimilable form, will not immediately adapt itself perfectly to a thin food like mothers' milk. If the entire mucous membrane is in a catarrhal condition, even breast milk itself would undergo ulterior change before it could be absorbed. It is a question of therapeutics as well as of food.

From the use of such empirical foods and empirical feeding has come the dictum, sometimes uttered "No food suitable for all cases—all foods must be tried." There is something superficially attractive about this proposition, but has it any place in a rational, scientific system of infant feeding? Too much insisted upon, does it not make any food good enough to sell? Is it not a palpably empirical standpoint? To what purpose then the comparative study and analysis of animal milk, and of the method of approximating it to the composition of human milk? Of what significance then, the theoretically accepted and unassailable postulate that mothers' milk is the standard of perfection?

Whatever part food in all its varieties plays in the therapeutics of infant feeding, there can be no escape from the logic of the proposition that the food practically identical with mothers' milk should be the food chosen for the artificial nourishment of an infant from birth.

Such a food is yielded by the Peptogenic Milk Powder and as such, it deserves the wide and general use so long given to empirical foods—to foods palpably unlike mothers' milk in physical characteristics and widely dissimilar in chemical composition.

It is often said that there are infants who will live on

anything and there are certainly also many with constitutions so feeble, so prone to disease that no care avails to succor. Whilst the empirical foods are used as broadcast as they are advertised and thus largely for average healthy infants, the "humanised milk" finds principal use in cases brought to the attention of the physician after failure with a variety of foods.

Peptogenic Milk Powder is the most successful food for sick and feeble infants, simply because it is the most like mothers' milk and it is the best food for healthy infants for the same reason. Why should the best food be selected only for the sick and feeble infant, the best food is the right food for the healthy infant also. "Just as the twig is bent the tree's inclined," and it is difficult to exaggerate the influence of proper feeding in the development and future health of the child.

RICH MILK—FROM ONE COW.

The formula and process for the preparation of "humanised milk" given, is based upon results obtained with ordinary, fresh cows' milk, as supplied by reputable dealers in all large cities. With this milk the best results are in practice obtained, both in the behavior of the milk with the Powder, and as a food. But we find that people are apt to obtain, when possible, the milk of rich Alderney cows, or one cow's milk. With this milk there is very apt to be trouble. It is not nearly so readily approximated to human milk as the ordinary mixed cows' milk. The richness of such milk is valuable when it is concerned in cheese making, but quite the contrary in preparing milk for a bottle fed infant. We believe that our views and our experience in this particular are in accordance with the best medical opinion at the present time.

CREAM.

The use of cream is not an indispensable condition to the employment of the Peptogenic Milk Powder. It is necessary with this as with *every other* food if we wish to get the amount of fat contained in human milk.

The use of cream is urgently advised. It should not be dispensed with upon an impression (for which there is no foundation in fact) that the cream is " too rich " for a child. A certain proportion of fat is provided in the natural food of an infant, and in a condition ready for absorption. It sustains important functions in the digestive process of an infant aside from that of nutrition.

"Skimmed milk" forms a peculiarly firm and tough curd—"hickory curd" as it is called. The presence of the cream undoubtedly aids the digestibility of milk, especially for infants. It gives mobility and softness to the curds, preventing the aggregation of large impenetrable masses. It is a significant fact that whilst every "infant food " sold, or the food as prepared with them, is deficient in cream, the use of *cream is not directed* except in the Fairchild process.

If it is found inconvenient to use cream, it is better to use the Peptogenic Milk Powder without cream than to resort to some other food, not only deficient in cream, but deficient and inferior in other respects also.

THE TEMPERATURE OF THE WATER BATH.

The object of the immersion in the water bath, in Directions No. 2, is to bring the milk (in the bottle) to about blood heat conveniently and without risk of over-heating. The water in the "bath" should be about 115° F.

The average temperature tolerable at which the *whole hand* can be immersed in water for *one* minute is about 115° F. It is seldom that any person can endure it more than a few degrees hotter. This expedient is, therefore, convenient and reliable for ascertaining the proper temperature of a vessel of water. Those who prefer may use the ordinary thermometer. "Dairy thermometers" or "bath thermometers," just the thing, may be purchased for a small sum.

The pail used for the hot water bath should hold a sufficient quantity of water to come up above the mixture in the bottle. It is not meant that the pitcher or pail of hot water containing the bottle of milk should be set in a warm place with the purpose of maintaining the same heat as started with. The water bath should stand in any convenient place at ordinary temperature of the room.

MILK TASTES BITTER.

"Humanised milk" properly prepared by the regular Directions No. 1 will not taste bitter; the milk may become bitter if it is too slowly heated to the boiling—as for instance, over a low fire. To avoid bitterness, in Directions No. 2, it is simply necessary to reduce time in water bath and to boil quickly.

In preparing the milk for cases of cholera infantum and cases of exceedingly feeble digestion, it is often desirable to digest the milk thirty minutes or so before boiling it. This milk (Directions No. 2) may taste bitter because of the very complete digestion of the caseine; but it is seldom refused by the infant, and if so, it may be sweetened with milk sugar, which may be bought of the druggist.

MILK CURDLED WHEN BOILED.

If a fine, granular curd appears in the "humanised milk" when it is mixed and boiled according to directions, it is because the milk is stale, or is too rich, or has not been mixed with the full amount of water. If the milk is fresh and is fit for use and has been properly diluted, it will not curd.

We find that people are apt to leave out the proper amount of water, because they think it is "too much."

THE USE OF CONDENSED MILK WITH PEPTOGENIC MILK POWDER.

Milk deprived of a definite proportion of water by evaporation, should theoretically become again like cows' milk in composition, by the addition of water. But in practice the condensed milks of commerce do not meet these anticipations. They cannot replace fresh whole milk as a basis for "humanised milk." We find no reason to recommend condensed milk when ordinary fresh milk is obtainable. Sweetened condensed milk should never be used;

it contains a large amount of cane sugar. The best condensed milk is usually not more than four times the strength of pure milk.

If it is absolutely necessary to use condensed milk, one part of pure unsweetened milk should be first mixed with from two and a half to three parts of water, and may then be presumed to be equivalent to cows' milk. Then to 8 ounces of this mixture add 8 ounces of water and the cream and Peptogenic Milk Powder in the usual manner. In other words, we first require to dilute pure, unsweetened condensed milk with about 7 parts of water, and to each *pint* of this diluted milk should be added 4 tablespoonfuls of cream and one large measure of the Peptogenic Powder and treated in the usual manner.

STERILISED MILK.

Sterilised milk has recently attracted much attention as a food for infants and as the basis of a food.

The sterilisation of milk has been advocated for the following reasons, viz. :—That milk in the udder contains no germs ; that the suckling is presumed to receive from a healthy source germless milk.

But normal sterile milk and sterilised milk differ in very significant particulars—one is a product of nature, the other accomplished by prolonged subjection to boiling under pressure.

Presuming that the process of sterilisation has been successfully conducted, and presuming that no other change has been produced by the sterilisation, theoretically sterilised milk precisely resembles pure, *germless* cows' milk.

But it is important to investigate the effects really produced upon cows' milk by the process of sterilisation, by boiling milk in a flask excluded from air for 30 to 45 minutes.

As a matter of fact, important changes are produced in the milk by this treatment. It is found that the amount of coagulable albumen is increased, that the milk is altered in its properties, and in its behavior and that these changes altogether indicate that milk is rendered more indigestible by sterilisation. The result of expert chemical investigation of sterilised milk at the present time may be said to be decidedly unfavorable to it as a substitute for breast milk.

By sterilisation, the ratio or sum of nutritive constituents of milk is not changed; its constituents are not in any degree brought to a greater resemblance to those of human milk.

Having, during the past few years, made careful and repeated examinations of sterilised milk, and with very considerable opportunities of becoming acquainted with the results of its practical use as a food for infants, we are unable to recommend sterilised milk. In fact, we have felt constrained to advise our correspondents against its use. One of the characteristic experiences reported has been that sterilised milk often proves incapable of affording adequate nourishment, especially for infants with impaired digestion. Such infants, although frequently fed with sterilised milk and in ample quantity, do not thrive, and evince a constant craving for food. Furthermore, the process for the use of Peptogenic Milk Powder, in which we recommend that the milk be brought simply to the boiling point, practically renders the milk sterile, free from germs and entirely suitable for the food of an infant for the length of time in which it is required, twenty-four hours or more. Even a much lower temperature, 160° to 170° F., is perfectly effective for all practical purposes; it kills the germs and kills the digestive ferment and consequently checks digestion. This temperature was proposed by Pasteur for the destruction of germs and the preservation of foods, and the process, as practically employed, is known as Pasteurisation. We have found that milk so prepared with the Peptogenic Powder and " Pasteurisation " will keep for 24 hours or more without change, simply corked in an ordinary bottle. We have very often had occasion to recommend this method as a substitute for sterilised milk, and with perfect success. There remains consequently no necessity whatever for submitting milk to the sterilising process. In this opinion, we believe that we have the concurrence of all experts, both chemists and physicians, who have given careful investigation to this subject.

HOBOKEN, N. J., June 14, 1884.

	Water.	Fat.	Milk Sugar.	Albuminoids.	Ash.
Average of Analyses 80 Samples of Woman's Milk.	86.73	4.13	6.94	2.	0.2
Analysis "Humanised Milk," as made with Peptogenic Milk Powder.	86.2	4.5	7.	2.	0.3

April 1st, 1891.

Messrs. FAIRCHILD BROS. & FOSTER.

DEAR SIRS :

It is now some seven years since I made my original report to you, in which I stated that I found the Peptogenic Milk Powder to yield a "humanised milk, which in taste, physical characteristics and chemical constitution approaches very closely to woman's milk."

During this time, I have at frequent intervals analysed the humanised milk as prepared with the Peptogenic Powder ; have made many analyses of milk and of "infant foods," and have studied the various methods of treating milk for the artificial feeding of infants. As a result of this experience, I feel confirmed in the conviction that the Peptogenic Milk Powder with the method given is the most exact, natural and practical means at present known of rendering cows' milk suitable as a comprehensive substitute for woman's milk.

Yours truly,

ALBERT R. LEEDS, PH.D.

Professor of Chemistry, Stevens Institute of Technology, Hoboken, N. J.

PRACTICAL RECIPES

FOR

PEPTONISED FOODS FOR THE SICK, MILK, GRUEL, BEEF, OYSTERS, JELLIES, PUNCHES, Etc.,

BY THE FAIRCHILD PROCESS.

These recipes are designed to facilitate the preparation of peptonised milk and other artificially digested foods. Their preparation requires only the simplest culinary utensils, and no more care or skill than that expended in making the ordinary foods for the sick, so long in vogue.

Of their infinite superiority, not only as *material* for *nutrition*, but in *adaptability for digestion* by the sick, it is scarcely necessary to speak.

Peptonised Foods are the chief reliance of the medical profession, both in private and hospital practice, for the feeding of the sick. And it is greatly to be hoped that the very fallacious ideas prevalent among the laity as to what constitutes a food for the sick, will, in spite of tradition and habit, give way to the more salutary and enlightened views now reached in the progress of medical science.

The subject of nutrition is now recognised to be of first importance in the treatment of disease. The sick require veritable food and digestible food. There are no "active principles" of food which can be extracted like alkaloids from drugs. By the Fairchild peptonising process foods which are found adequate for the nourishment of the healthy and vigorous may be adjusted to the functions of digestion enfeebled by chronic ailments, or wholly interrupted by acute diseases, fevers, etc.

THE NUTRITIVE VALUE OF MILK AS COMPARED WITH BEEF TEA, EXTRACTS OF BEEF, ETC.

Milk contains sugar ready formed for absorption ; fat in a condition perfectly adapted for assimilation ; mineral substances essential to nutrition of the bony structure ; and a due proportion of albumen, or flesh forming element—caseine.

One pint of milk contains over two ounces of actual dry, solid nutritious substance.

" Beside the trifling amount of proteid material, and the fat (which lat-
" ter is guarded against with great care) the beef tea then only contains
" the salts of the muscle, the hematin and allied pigments, traces of
" sugar perhaps, some lactic acid, and the nitrogenous extractives, creatin
" and its congeners.

" As the original half pound of muscle will contain but forty to sixty
" grains of salt, and ten to twelve of nitrogenous waste products, the
" beef tea certainly contains no more."

Prof. BAUMGARTEN, M.D.

" The valuation by most persons outside the medical profession, and
" by many within it, of beef tea or its analogues, the various solutions,
" most of the extracts and the expressed juice of meat, is a delusion and
" a snare which has led to the loss of many lives by starvation. The
" quantity of nutritive material in these preparations is insignificant or
" nil, and it is vastly important that they should be reckoned as of little
" or no value, except as conducive indirectly to nutrition by acting as
" stimulants for the secretion of the digestive fluids or as vehicles for
" the introduction of nutritive substances. Furthermore, it is to be con-
" sidered that water and pressure not only fail to extract the alimentary
" principles from meat, but the excrementitious principles, or the prod-
" ucts of destructive assimilation, are thereby extracted. A few years
" ago, a German experimenter declared that he produced fatal toxæmia
" in dogs by feeding them with this popular article of diet."

Dr. AUSTIN FLINT, Sr.

So much then for this " strength " that so many people fancy they get out of the beef, by the maceration in cold water, simmering and boiling. How much less does the beef weigh than at the beginning ?

It is the flesh that gives value to the beef, wherein it differs from farinaceous foods. The flesh is *not* soluble in water. The water extracts some of the salts of the beef, some coloring matter, extractives, etc., and the now tasteless flesh is discarded. Beef tea, beef extract, is utterly incapable of properly nourishing the body in health or disease. Milk does supply every element of nutrition, the elements that are found in the most diverse forms of food.

RECIPES.

PEPTONISED MILK.

WARM PROCESS.

Into a clean quart bottle put the powder contained in one of the peptonising tubes, and a teacupful of cold water, shake, then add a pint of fresh cold milk and shake the mixture again. Place the bottle in water so hot that the whole hand can be held in it without discomfort for a minute (or at about 115° F.).

Keep the bottle there ten minutes.

At the end of that time put the bottle on ice to check further digestion and keep the milk from spoiling.

Place the bottle directly in contact with the ice.

Ten minutes in the hot water-bath gives sufficient time for the predigestion of the milk in ordinary cases.

If there is any evidence that the milk requires more digestion, it is only necessary to let the milk stand a longer time in the hot water-bath.

COLD PROCESS.

Mix the peptonising powder in cold water and cold milk, as usual, and immediately place the bottle on ice, without subjecting it to the water-bath or any heat.

When needed pour out the required portion, and use in the same manner as ordinary milk.

It is recommended to try the milk prepared by the COLD process, in those cases in which food is not quickly rejected after ingestion, but in which the digestive functions are impaired, or even practically suspended. It has been found in many such cases that the peptonising principle exerts sufficient action upon the milk in the stomach to insure its digestion and proper assimilation. If the milk so prepared be not well borne, or any evidence appear of its imperfect digestion, it should be sufficiently predigested—peptonised—by the usual warm process.

Milk by the "cold process" is especially suited for dyspeptics and persons who ordinarily find milk indigestible. This milk has no taste or evidence of the presence of the peptonising agent.

PARTIALLY PEPTONISED MILK.

Put into a clean granite ware or porcelain lined saucepan the powder contained in one of the Fairchild peptonising tubes, and a teacupful (gill) of cold water ; stir well, then add a pint of fresh cold milk. Place the saucepan on a hot range or gas stove and heat with constant stirring until the mixture boils. The heat should be so applied as to make the milk boil in ten minutes. When cool, strain into a clean bottle, cork well and keep in a cool place. When needed, shake the bottle, pour out the required portion, and serve cold or hot as directed by the physician in charge.

N. B.—Milk thus prepared will not become bitter.

HOT PEPTONISED MILK, AS A BEVERAGE.

Into a clean quart bottle put the powder contained in one of the Peptonising Tubes, and a teacupful of cold water, shake, then add a pint of fresh cold milk and shake the mixture again. Place the bottle on ice until the milk is required for use. When needed, pour the portion to be used into a saucepan and heat as hot as can be agreeably sipped.

If required for immediate use, the peptonising powder, cold water and cold milk may be thoroughly mixed in the saucepan and heated to the proper temperature for drinking.

At this temperature (during the heating) the peptonising powder acts with great rapidity, and in a few minutes a hot peptonised milk may be prepared which will be sufficiently digested for the majority of cases.

Hot peptonised milk is the most grateful, nourishing and bracing beverage for invalids, dyspeptics, diabetics and consumptives.

It is especially useful with breakfast, and at any time when suffering from a sense of exhaustion with an intolerance for solid foods.

It is very acceptable to persons who require nourishment before sleeping and may be used at the table instead of ordinary milk with tea or coffee.

EFFERVESCENT PEPTONISED MILK.

Put some finely cracked ice in a glass and then half fill it with cold apollinaris, vichy, clysmic or carbonic water as preferred, then quickly pour in the peptonised milk and drink during effervescence.

Peptonised milk may be made agreeable to many patients by serving with a little grated nutmeg, sweetened, or flavored with a little brandy, etc.

SPECIALLY PEPTONISED MILK.

FOR JELLIES, PUNCHES, ETC.

FOR ALL RECIPES WHERE THE MILK IS TO BE MIXED WITH FRUIT JUICES OR ACIDS.

Mix the peptonising powder, water and milk, in a bottle, and place in a hot water-bath exactly as directed in the warm process recipe. Now let the bottle remain in the hot water for *one hour*, then pour into a saucepan and HEAT TO BOILING. This specially peptonised milk is now ready for use in making jellies, etc. It may be immediately used if required hot, or set aside on ice for punches, etc.

In peptonising milk for all these recipes in which lemon juice or acid is to be used, it is necessary to carry the process to the point at which the milk will not curdle with acid. Hence the one hour digestion.

Do not fail to boil the milk immediately after the one hour in water-bath in order to kill the peptonising ferment which would otherwise digest the gelatine when added and thus prevent the milk from forming a jelly.

The bitter taste of the milk so peptonised, is entirely absent from the jellies, punches, etc., and these foods containing milk in a completely digested form are not only agreeable, but exceedingly assimilable.

PEPTONISED MILK JELLY.

First take about half a box of Cox's Gelatine and set it aside to soak in a teacupful of cold water until needed.

Take one pint of hot "*specially*" peptonised milk and dissolve in it about a quarter of a pound of sugar, or sufficient to taste, next add the gelatine and stir until dissolved.

Pare one fresh lemon and one orange, and put the rinds into the hot peptonised milk.

Squeeze the lemon and orange juice into a glass, strain, and mix it with two or three tablespoonfuls of best St. Croix Rum, or brandy, etc., as may be preferred.

Lastly add the juices and the spirits with stirring.

Strain all through a colander and when cooled to a syrup consistency, so as to be almost ready to " set," pour into tumblers or jelly moulds and put in a cold place.

It is important not to pour the milk into the moulds until it is nearly cool, otherwise it will separate in setting.

This jelly has a delicious flavor, is highly acceptable to invalids and convalescents at the period when they tire of liquids and crave more substantial food.

Good St. Croix Rum is generally preferable to other spirits in making jellies, punches, etc.

PEPTONISED MILK PUNCH.

Prepare a punch from peptonised in the same manner as from or-

dinary milk, using St. Croix or Jamaica Rum, Whiskey or Brandy as preferred, and served with grated nutmeg.

This is a good way :

Take a goblet about one-third full of fine crushed ice, pour on it a
tablespoonful of St. Croix Rum, a dash of Curaçoa, or other liquor that
is agreeable to the taste, then fill the glass with peptonised milk, stirring
well, sweeten to taste, grate a little nutmeg on top.

PEPTONISED MILK LEMONADE.

Take a goblet one-third full of cracked ice, squeeze on it the juice of a
lemon, and dissolve sufficient sugar, then fill the glass with *specially* peptonised milk, stirring well.

Make this lemonade of equal parts of peptonised milk and mineral
water, instead of milk alone, if you prefer, first pouring the water, lemon
juice, etc., on the ice, and then filling the glass with the milk.

This makes an effervescing punch that is very agreeable.

PEPTONISED MILK GRUEL.

Mix smoothly a heaping teaspoonful of wheat flour or arrowroot, with
half pint of cold water. Then heat with constant stirring until it has
boiled briskly for several minutes.

Mix with this hot gruel one pint of cold milk and *strain* into a small
pitcher or jar, and immediately add the contents of one "peptonising
tube," mix well. Let it stand in the hot water-bath, or warm place, for
20 minutes, then put in a clean quart bottle and place on ice.

This milk gruel may be used in the same manner and for the same purpose as plain peptonised milk.

The flavor of this milk gruel is very agreeable ; the taste of the peptone being masked by the digested arrowroot or flour, the peptonising
powder digesting both the farinaceous matter and the milk.

PEPTONISED MILK WITH PORRIDGE.

To a dish of porridge of oatmeal, rice, hominy, etc., as prepared
for the table, add a sufficient quantity of hot or cold peptonised
milk.

It will aid in the digestion of farinaceous foods for young children, as
well as supplying the milk in a form especially adapted for children with
defective digestion.

PEPTONISED BEEF.

Take one-quarter pound finely minced, raw *lean* beef, or same weight (of equal portions) of beef and chicken meat mixed.

Cold water, half a pint.

Cook over a gentle fire, stirring constantly until it has boiled a few minutes.

Then pour off the liquor, for future use, and beat or rub the meat to a paste, and put it into a clean fruit jar or bottle with half a pint of cold water and the liquor poured from the meat.

Add—

Extractum Pancreatis.........4 measures	(20 grains).
Soda Bicarb.................1 measure	(15 grains).

Shake all well together, and set aside in a warm place, at about 110° to 115°, for three hours, stirring or shaking occasionally ; then boil quickly.

It may then be strained, or clarified with white of egg, in usual manner. Season to taste with salt and pepper.

For great majority of cases it will not be required to strain the peptonised liquor, for the portion of meat remaining undissolved will have been so softened and acted upon, by the pancreatic extract, that it will be in very fine particles and diffused in an almost impalpable condition. Thus in a form readily subject to digestion in the stomach.

FARINACEOUS materials may also be advantageously used in the preparation of the peptonised soup, by simply boiling a sufficient quantity of flour, arrowroot, etc., with a *half portion* of the water used in above recipe, and mixing all together—meat, gruel, Extractum Pancreatis and Soda. The Extractum Pancreatis will, at the same time, digest both *starch* and *meat*.

This has a more agreeable flavor than that made of meats alone.

Jelly also may be made of peptonised beef.

Be sure to boil the peptonised beef, after three hours in warm place, otherwise the digestion will progress until it is spoiled.

PEPTONISED OYSTERS.

(Originally suggested by DR. N. A. RANDOLPH)

Take half a dozen large oysters with their juice and half a pint of water. Heat in a saucepan until they have boiled briskly for a few minutes. Pour off the broth and set aside.

Mince the oysters finely, and reduce them to a paste with a potato masher in a wooden bowl.

Now put the oysters in a glass jar with the broth which has been set aside and add

Extractum Pancreatis..........3 measures   (15 grains).

Soda Bicarb ....1 measure   (15 grains).

Let the jar stand in hot water or a warm place where the temperature is not above 115 degrees, for one and a half hours.

Then pour into a saucepan and add half a pint of milk.

Heat over the fire slowly to boiling point.

Flavor with salt and pepper, or condiments, to taste and serve hot.

There will be found but very small bits of the oysters undigested, and these may be strained out or rejected in eating the soup, but will not be unacceptable to the stomach, except in very rare cases.

The *milk* will be sufficiently digested during the few minutes which will elapse before the mixture boils, if heated gradually.

Be sure to boil the peptonised oysters to finish the process.

JUNKET, OR CURDS AND WHEY,

WITH

FAIRCHILD'S ESSENCE OF PEPSINE.

Junket, the soft jelly-like curded milk as prepared with Fairchild's Essence of Pepsine, is a delicious delicacy for invalids, convalescents and dyspeptics. It is especially acceptable and appropriate in *convalescence*, when the liquid foods have become tiresome and repulsive. This junket gives the grateful and wholesome sense of substance, whilst it does not oppress the digestion.

Take half a pint of fresh milk heated lukewarm, add one teaspoonful of Essence of Pepsine, and stir just enough to mix. Pour into custard cups, let it stand till firmly curded ; may be served plain or with sugar and grated nutmeg.

AS A DESSERT, junket when served with cream, sweetened and flavored with nutmeg or wine, is far more toothsome than more elaborate dishes and has the merit of requiring but a few minutes and no special skill in its preparation.

JUNKET OF MILK AND EGG,

MADE WITH

FAIRCHILD'S ESSENCE OF PEPSINE.

Beat one egg to a froth and sweeten with two teaspoonfuls of white sugar, add this to half a pint of warm milk ; then add one teaspoonful of

Essence of Pepsine, let it stand till curded.   This milk and egg junket is a highly nutritious and agreeable food.

WHEY

MADE WITH

FAIRCHILD'S ESSENCE OF PEPSINE.

Take half a pint of fresh milk heated lukewarm, (about 115° F) add one teaspoonful of Essence of Pepsine and stir just enough to mix ; when firmly curded, beat up with a fork until the curd is finely divided, now strain and the Whey is ready for use.   Whey contains in solution the soluble albuminoids, the sugar and the salts (mineral constituents) of the milk and a small portion of fat.

It is therefore a nutritious fluid food peculiarly useful in many ailments and always valuable as a means of variety in diet for the sick. It is frequently resorted to as a food for infants to tide over periods of indigestion, summer complaints, etc.   Whey is in some cases indicated with wine or brandy and may then be mixed with the spirit.

Whey, or curds and whey, as made with *Fairchild's Essence of Pepsine* is superior to that made with liquid rennet, because of the peptic as well as the curdling activity of the *Essence*, and is moreover far more acceptable to the stomach.   (For use of Whey in Cholera Infantum see pages 96-97.)

THE PARTIAL DIGESTION OF FARINACEOUS FOODS AT THE TABLE.

To a saucer of well-cooked porridge of oatmeal, wheaten grits or rice, etc., as warm as proper to be eaten, add one to two teaspoonfuls *Diastasic Essence of Pancreas*.   Stir for a few minutes until thoroughly mixed, before eating it.

The Diastasic Essence must not be added to very hot food, for if hotter than can be agreeably borne by the mouth, the digestive principle will be destroyed.

Extractum Pancreatis may be added in exactly the same manner, using a measure full of the dry Extractum Pancreatis instead of the teaspoonful of Diastasic Essence.   The powder imparts no taste or odor to the food and is handy to use.   It further contains every digestive principle— those capable of digesting milk, fat, etc., and thus will aid in the digestion of the ordinary foods taken at the same meal with the porridge.

FAIRCHILD'S PREPARATIONS.

Pepsin in Scales,

Pepsin in Powder,

Essence of Pepsine,

Saccharated Pepsin,

Glycerinum Pepticum,

Extractum Pancreatis,

Diastasic Essence of Pancreas,

Peptonising Tubes,

Peptogenic Milk Powder,

Panopepton,

Pancreatic Tablets,

Compound Pancreatic Tablets,

Pepsin and Extract Pancreatis Tablets,

Pepsin and Bismuth Tablets,

Pepsin, Bismuth and Pancreatic Tablets,

Pepsin, Bismuth and Nux Vom. Tablets,

Pepsin and Diastase Tablets,

Peptonate of Iron Tablets,

Compound Ox Gall Tablets,

Ferroglobin Tablets,

Trypsin.